Beginners Guide
To Observing The Constellations

Errol Jud Coder

Written and Designed by Errol Jud Coder

Constellation drawings by Johannes Hevelius & Johann Bode in their individual books titled "**Uranographia**" published before 1923 and public domain in the US.

Constellation Star charts by IAU and Sky & Telescope magazine (Roger Sinnott & Rick Fienberg) licensed under the Creative Commons Attribution 3.0 Unported license. http://creativecommons.org/licenses/by/3.0/legalcode

Dreamscape Publishing
http://dreamscapepublishing.webs.com

© 2014 Errol Jud Coder

No part of this book may be reproduced in any form for any electronic or mechanical means, including information storage and retrieval devices or systems, without prior written permission, except for those in public domain, and others, brief passages may be quoted for reviews.

Printed in the United States
10 9 8 7 6 5 4 3 2 1

September 2014

ISBN-13: 978-1501069567

ISBN-10: 150106956X

Astronomy: ,
Adult, Young Adult & Juvenile

NOTE Every effort has been taken to ensure that all information in this book is correct and accurate. Contact us with any mistakes that you might find.

Table of Contents

Introduction............3
  Northern Circumpolar Constellations..........5
  Northern Hemisphere Constellations..........6
  Southern Hemisphere Constellations..........7
Reading Charts......8
Stellar Classifications.....16
  Luminosity Classes..................16
  Spectral types........17
Andromeda..........20
Antlia..................26
Apus....................28
Aquarius..............32
Aquila..................38
Ara......................40
Auriga.................48
Bootes.................52
Camelopardalis...62
Cancer.................68
Canes Venatici.....74
Canis Major.........78
Capricornus.........86
Carina..................90
Cassiopeia...........96
Centaurus..........102
Cepheus..............108
Cetus..................116
Chamaeleon.......120
Circinus..............122
Columba............124
Coma Berenices 126
Corona Australis 134
Corona Borealis 138
Corvus................142
Crater.................146
Crux...................150
Cygnus...............156
Delphinus..........162
Dorado...............166
Draco.................170
Equuleus............178
Eridanus.............182
Fornax...............186
Gemini..............190
Grus...................196
Heracles............200
Horologium......206
Hydra................208
Hydrus..............216
Indus.................218
Lacerta.............220
  Binary stars:........222
Leo Major.........224
Leo Minor........232
Lepus................236
Libra.................240
Lupus................244
Lynx..................248
Lyra..................252
Mensa...............258
Microscopium...260
Monoceros........262
Musca...............270
Norma..............272
Octans...............276
Ophiuchus.........278
Orion.................284
Pavo..................292
Pegasus.............296
Perseus.............300
Phoenix............306
Pictor................310
Pisces................316
Piscis Austrinus.320
Puppis...............324
Pyxis.................328
Reticulum.........330
Sagitta..............332
Sagittarius.........336
Scorpius............344
Sculptor............350
Scutum.............354
Serpens Caput...358
Sextans.............362
Taurus...............364
Telescopium......372
Triangulum........374
Triangulum Australe............376
Tucana..............378
Ursa Major........383
Ursa Minor.......393
Vela...................397
Virgo.................403
Volans...............411
Vulpecula..........413
Glossary............415
Bibliography.....416

1

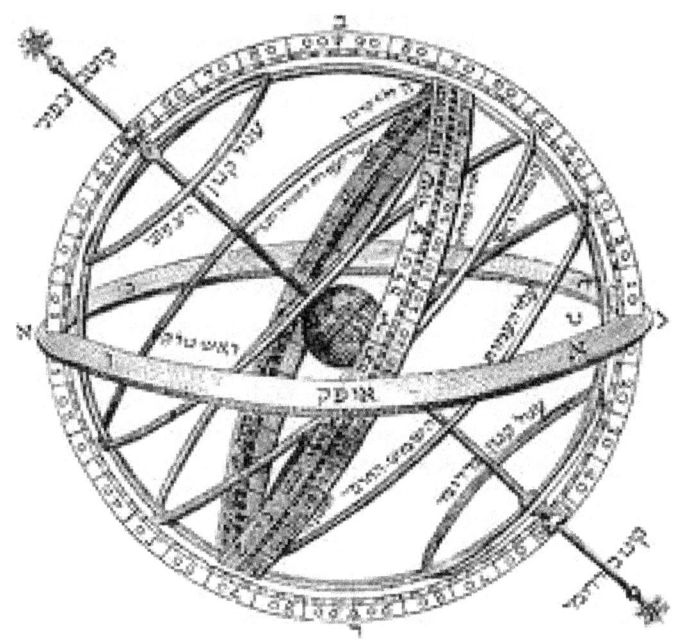

INTRODUCTION

Most ancient cultures saw pictures in the stars of the night sky. The earliest known efforts to catalog the stars date to cuneiform texts and artifacts dating back roughly 6000 years. These remnants, found in the valley of the Euphrates River, suggest that the ancients observing the heavens saw the lion, the bull, and the scorpion in the stars. The constellations as we know them today are undoubtedly very different from those first few--our night sky is a compendium of images from a number of different societies, both ancient and modern. By far, though, we owe the greatest debt to the mythology of the ancient Greeks and Romans.

The earliest references to the mythological significance of the Greek constellations may be found in the works of Homer, which probably date to the 7th century B.C. In the Iliad, for instance, Homer describes the creation of Achilleus's shield by the craftsman god Hephaistos:

> On it he made the earth, and sky, and sea, the weariless sun
> and the moon waxing full, and all the constellations
> that crown the heavens, Pleiades and Hyades, the mighty

Orion and the Bear, which men also call by the name of Wain: she wheels round in the same place and watches for Orion, and is the only one not to bathe in Ocean (Iliad XVIII 486-490).

At the time of Homer, however, most of the constellations were not associated with any particular myth, hero, or god. They were instead known simply as the objects or animals which they represented--the Lyre, for instance, or the Ram. By the 5th century B.C., however, most of the constellations had come to be associated with myths, and the Catasterismi of Eratosthenes completed the mythologization of the stars. " At this stage, the fusion between astronomy and mythology is so complete that no further distinction is made between them"

--the stars were no longer merely identified with certain gods or heroes, but actually were perceived as divine (Seznec, 37-40).

Constellations are formed of bright stars which appear close to each other on the sky, but are really far apart in space. The shapes you see all depend on your point of view. Many societies saw patterns among the stars with gods and goddesses or stories from their cultures.
Many peoples noticed that the planets, the moon, and comets moved through the sky in a different way than the stars.

Because of the rotation of the Earth and its orbit around the Sun, we divide the constellations into two groups. Some constellations never rise nor set, and they are called *circumpolar*. All the rest are divided into *seasonal* constellations. Which constellations will be circumpolar and which seasonal depends on your latitude.

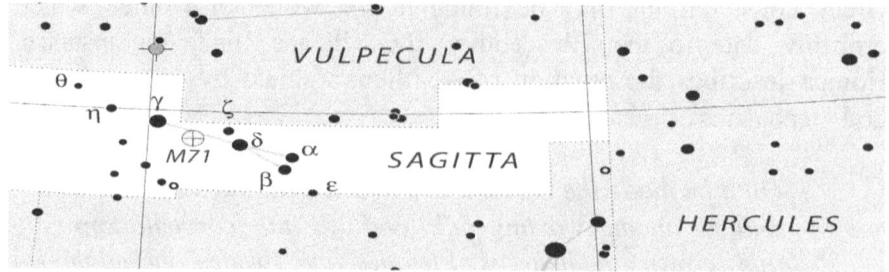

Northern Circumpolar Constellations

Because of the rotation of the Earth and its orbit around the Sun, we divide the stars and constellations into two groups. Some stars and constellations never rise nor set, and they are called circumpolar. All the rest are divided into seasonal stars and constellations. Which stars and constellations will be circumpolar and which seasonal depends on your latitude. In the northern hemisphere, we will always be able to see stars and constellations in the the northern circumpolar sky, while in the southern hemisphere, we will always be able to see stars and constellations in the southern circumpolar sky.

Constellations in the northern circumpolar sky include Auriga, Camelopardalis, Cassiopeia, Cepheus, Draco, Lynx, Perseus, Ursa Major, and Ursa Minor. These constellations are always visible in the night sky of the Northern Hemisphere.

Constellations in the southern circumpolar sky include Grus, Phoenix, Indus, Tucana, Pavo, Ara, Eridanus, Hydrus, Horologium, Reticulum, Octans, Apus, Triangulum Australe, Lupus, Circinus, Musca, Crux, Centaurus, Carina, Vela, Puppis, Dorado, and Chamaeleon. These constellations are always visible in the night sky of the Southern Hemisphere.

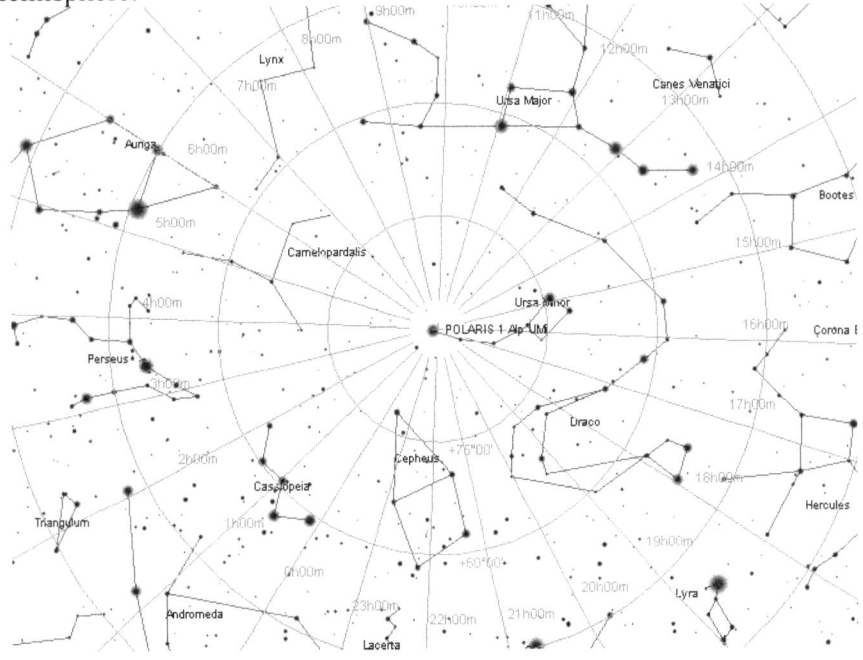

Northern Hemisphere Constellations

Many different constellations fill the evening sky in the northern hemisphere. Depending on your location and the season, different constellations can be seen. Northern circumpolar constellations can be seen all year long in the night sky of the northern hemisphere, and appear to circle about the Pole star. This image shows an illustration of Ursa Major, the Great Bear.

Northern Circumpolar Constellations	Northern Spring Constellation	Northern Summer Constellations	Northern Autumn Constellations	Northern Winter Constellations
Cassiopeia	Bootes	Aquila	Andromeda	Canis Major
Cepheus	Cancer	Cygnus	Aquarius	Cetus
Draco	Crater	Hercules	Capricornus	Eridanus
Ursa Major	Hydra	Lyra	Pegasus	Gemini
Ursa Minor	Leo	Ophiuchus	Pisces	Orion
	Virgo	Sagittarius		Perseus
		Scorpius		Taurus

Southern Hemisphere Constellations

Many different constellations fill the evening sky in the southern hemisphere. Depending on your location and the season, different constellations can be seen. Southern circumpolar constellations can be seen all year long in the night sky of the southern hemisphere. In the southern hemisphere, there is no bright pole star. This image shows an illustration of Crux - the Southern Cross.

Southern Circumpolar Constellations	Southern Spring Constellations	Southern Summer Constellation	Southern Autumn Constellations	Southern Winter Constellations
Carina	Andromeda	Canis Major	Bootes	Aquila
Centaurus	Aquarius	Cetus	Cancer	Cygnus
Southern Cross	Capricornus	Eridanus	Crater	Hercules
	Pegasus	Gemini	Hydra	Lyra
	Pisces	Orion	Leo	Ophiuchus
		Perseus	Virgo	Sagittarius
		Taurus		Scorpius

Reading Charts

Looking at the charts with out knowing what everything means can become overwhelmingly hectic. A basic understanding of what you are seeing will aide you in a better usages, so you can get the best out of it. The best way to understanding what you are seeing, and knowing how to use it, is to take it step by step.

Lets begin with by using the Andromeda constellation chart as an example.

Constellation fields

Each of the 88 constellations are given an designated area in the sky. On the charts the white area corresponds to the particular constellation being discussed. The darker areas are then everything outside of the constellation field.

Magnitude

Lets first begin with understanding the size of the stars you see on the charts. The Magnitude, which is the apparent brightness of the objects a human sees in the sky are denoted by their particular size on the chart. The bigger a dot is, means it is that much brighter. The smaller it gets, the dimmer it is. Typically, the human eye can only see as low as a 6-7 magnitude star. So, with this in mind the faintest objects shown in the charts will be down to 6 magnitude. You will find the magnitude legend at the bottom left of each chart.

Stars, Nebula & Galaxies

The objects on the chart represent different celestial bodies.

The **solid dots** correspond to stars, size determining their magnitude.

The **dark ovals** correspond to galaxies.

The **light circles** correspond to nebula, globular clusters, and other similar objects.

Many of these objects are given some sort of designation, whether it is a Greek alphabet character, a number, name, or a letter. Those that are shown with their common name also have a number designation based on the type of object it is. The list of common objects observable by the naked eye will include some of these designations themselves.

What are the numbers bordering the chart?

Chart Coordinates

If you have a list of locations and you want to find them, you need a way of doing so. Just like you can find any location on the Earth using Latitude and Longitude coordinates, you can do the same looking outward into space. With the Earth Lat and Long lines are overlayed on Earth based charts in vertical and horizontal lines that encircle the globe/sphere of earth.

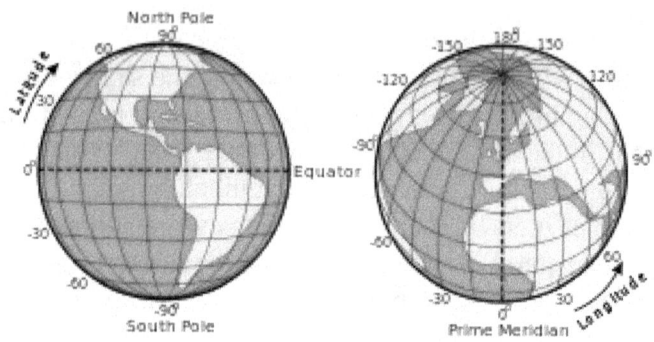

     The Longitude lines encircle the earth and range between 0 to 24hr segments (h). In between each hour segment, they are then separated by divisions of 60 minutes (m) and even smaller, divisions of 60 seconds (s). This corresponds to the time the Sun passes over the earth as the Earth orbits. It takes 24 hours as the earth spins, so it can be split up by 24 sections. When it comes to pin pointing a location though, the hours, minutes and seconds do not represent time, just points along the coordinate system. 0 hours starts at GMT/UTC. The line passes from the North Pole, through Greenwich England then to the South Pole, then moves West to East for 24 hour divisions.

     The Latitude are North to South lines divided up by degrees (°). The Equator (the center of our rotation) marks the 0 degree line and goes north to the North Pole which equals 90 degrees, and going south to the South Pole to -90 degrees. Degrees can be separated even smaller by 60 minutes (') of arc for each degree, and 60 seconds (") of arc for each minute of arc, allowing pin pointing of a location.

Equatorial coordinate system

If you are given a the Lat and Long of any point on the Earth, you can find it on any chart.
Using this Earth based concept of a coordinate system, you can apply it to the **Equatorial coordinate system** of the Celestial sphere.

In astronomy and navigation, the **celestial sphere** is an imaginary sphere of arbitrarily large radius, concentric with Earth. All objects in the observer's sky can be thought of as projected upon the inside surface of the celestial sphere, as if it were the underside of a dome.

This is the celestial globe. The imaginary equator of this globe is called the celestial equator. The imaginary path that the sun traces across the celestial globe during one day is called the Ecliptic. The constellations behind this path are called the Zodiac. While sun appears to move across 12 of them during a 24 hour period, in fact it is actually the Earth that is rotating. It only appears that the sun is the one moving.

The *equatorial* **coordinate system** is centered at Earth's center, but fixed relative to distant stars and galaxies. The coordinates are based on the location of stars relative to Earth's equator if it were projected out to an infinite distance. The equatorial describes the sky as seen from the solar system, and modern star maps almost exclusively use equatorial coordinates.

The *equatorial* system is the normal coordinate system for most professional and many amateur astronomers having an equatorial mount that follows the movement of the sky during the night. Celestial objects are found by adjusting the telescope's or other instrument's scales so that they match the equatorial coordinates of the selected object to observe.

Along the edge of the charts in this book are the right ascension and declination numbers that are used in this system. Using this system you can get the coordinates of any object and find it on any chart using the equatorial coordinate system.

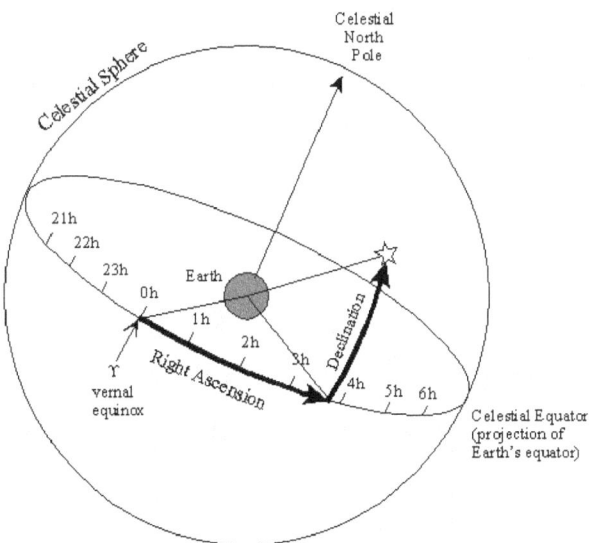

The stars on the celestial sphere are like cities on the globe. Cities are located on the globe using latitude and longitude. Longitude says how far the city is east or west *along* the Earth's equator; latitude says how far a city is north or south *of* the Earth's equator.

Right ascension *(RA)* is like longitude. It locates where a star is along the celestial equator. The zero point of longitude has been chosen to be where the line straight down from the Greenwich Observatory in England meets the equator. The zero point for right ascension is the vernal equinox. To find the right ascension of a star follow an hour circle "straight down" from the star to the celestial equator. The angle from the vernal equinox eastward to the foot of that hour circle is the star's right ascension.

There is one oddity in right ascension: the unit used to report the angle. Right ascensions are always recorded in terms of hours, minutes, and seconds. One hour of right ascension (1^h) is $15°$. Since $24 \times 15° = 360°$, there are 24^h of right ascension around the celestial equator. The reason for this oddity is that the celestial sphere makes one full rotation (24^h of RA) in one day (24 hours of time). Thus the celestial sphere advances about 1^h of RA in an hour of time.

Declination *(DEC)* is like latitude. It reports how far a star is from the celestial equator. To find the declination of a star follow an hour circle "straight down" from the star to the celestial equator. The angle from the star to the celestial equator along the hour circle is the star's declination, and is measured above the equator as 0 to 90 degrees (°), at the celestial north pole and is measured below the equator as 0 to -90 degrees (°) at the celestial south pole. Smaller units of degrees are separated by 60 minutes of arc (') and even smaller, 60 arcs of seconds (").

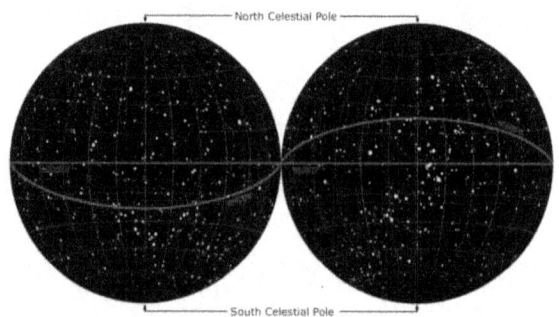

The Greek Alphabet

The Greek alphabet is commonly used in science, specifically astronomy for many aspects of nomenclature and designations of objects. The table below is to allow you to know which symbol corresponds to which character. It should help you better understand the different designations listed in this book and allow you to know which objects you are looking at.

A	α	alpha	N	ν	nu
B	β	beta	Ξ	ξ	ksi
Γ	γ	gamma	O	o	omicron
Δ	δ	delta	Π	π	pi
E	ε	epsilon	P	ρ	rho
Z	ζ	zeta	Σ	σς	sigma
H	η	eta	T	τ	tau
Θ	θ	thea	Y	υ	upsilon
I	ι	iota	Φ	φ	phi
K	κ	kappa	X	χ	chi
Λ	λ	lambda	Ψ	ψ	psi
M	μ	mu	Ω	ω	omega

The Usage of Time

UT is an Abbreviation for Universal Time

UT is used as a basis for calculating time throughout most of the world. UT is also called Greenwich Time, Greenwich Mean Time, Zulu Time. It is the time along the prime meridian (0 longitude) that runs through the Greenwich Observatory outside of London, UK, where the current system originated.

UT is tied to the rotation of the Earth in respect to the fictitious "mean Sun," which gives equal 24-hour days throughout the year. Greenwich Mean Time (GMT) was measured from Greenwich Mean at midday until 1925, when the reference point was changed from noon to midnight and the name changed to "Universal Time."

UT, GMT and UTC can be used interchangeably (*)

Using the Table Below

UT conversion is fairly easy. Just add a specified number of hours to UT time to determine your time. All you need to know is your time zone and if / when your state or country enforces Daylight Saving Time. If it does, you add an additional hour to Universal Time during that time of the year.

Take UT and Add
the following to determine your time:

Time Zone	Standard Time	Daylight Saving & Summer Time
Chatham Islands (CHAST) (CHADT)	+12.75 hrs	+13.75 hrs
International Date Line East (IDLE)	+12 hours	
New Zealand Standard Time (NZST) (NZDT)	+12 hours	+13 hours
Australian Eastern Standard Time (AEST) (AEDT)	+10 hours	+11 hours
Australian Central Standard Time (ACST)	+9.5 hours	+10.5 hours

(ACDT)

Japan Standard Time (JST)	+9 hours	
Australian Western Standard Time (AWST)	+8 hours	
India Standard Time (IST)	+5.5 hours	
Russian Zone 3	+4 hours	
Baghdad Time (BT)	+3 hours	
Moscow Time(MSK) (MSD)	+3 hours	+4 hours
Russian Zone 1	+2 hours	
Central European Time (CET)	+1 hour	
Middle European Time (MET)	+1 hour	
Swedish Winter Time (SWT)	0 hours	
Greenwich Mean Time (GMT)	0 hours	
Universal Time (UT)	0 hours	
Western European Time (WET) (WEST)	0 hours	+1 hour
West African Time (WAT)	-1 hour	
Newfoundland Standard Time (NST) (NDT)	-3.5 hours	-2.5 hours
Atlantic Standard Time (AST)(ADT)	-4 hours	-3 hours
Eastern Standard Time (EST)(EDT)	-5 hours	-4 hours
Central Standard Time (CST)(CDT)	-6 hours	-5 hours
Mountain Standard Time (MST)(MDT)	-7 hours	-6 hours
Pacific Standard Time (PST)(PDT)	-8 hours	-7 hours
Alaskan Standard Time (AKST)(AKDT)	-9 hours	-8 hours
Hawaiian Standard Time (HST)	-10 hours	

Stellar Classifications

In astronomy, **stellar classification** is the classification of stars based on their spectral characteristics. Light from the star is analyzed by splitting it with a prism or diffraction grating into a spectrum exhibiting the rainbow of colors interspersed with absorption lines. Each line indicates an ion of a certain chemical element, with the line strength indicating the abundance of that ion. The relative abundance of the different ions varies with the temperature of the photosphere. The *spectral class* of a star is a short code summarizing the ionization state, giving an objective measure of the photosphere's temperature and density.

Most stars are currently classified under the Morgan–Keenan (MKK) system using the letters *O, B, A, F, G, K, M, L, T* and *Y*, a sequence from the hottest (*O* type) to the coolest (*Y* type). The types *R* and *N* are carbon-based stars, the type *S* is zirconium-monoxide-based stars. Each letter class is then subdivided using a numeric digit with *0* being hottest and *9* being coolest (e.g. A8, A9, F0, F1 form a sequence from hotter to cooler).

Luminosity Classes

In the MKK system a luminosity class is added to the spectral class using Roman numerals. This is based on the width of certain absorption lines in the star's spectrum which vary with the density of the atmosphere and so distinguish giant stars from dwarfs. Luminosity class *0* stars for *hypergiants* , class *I* stars for *supergiants*, class *II* for bright *giants*, class *III* for regular *giants*, class *IV* for *sub-giants*, class *V* for *main-sequence stars*, class *VI* for *sub-dwarfs*, class *VII* for *white dwarfs*, and class *VIII* for *brown dwarfs*. The full spectral class for the Sun is then G2V, indicating a main-sequence star with a temperature around 5,800K.

A number of different **luminosity classes** are distinguished:

- **0**, Ia-0, Ia$^+$ hypergiants or extremely luminous supergiants (later addition)), Example: Eta Carinae (spectrum-peculiar)
- **Ia** (luminous supergiants), Example: Deneb (spectrum is A2 Ia)
- **Iab** (intermediate luminous supergiants), Example: Betelgeuse

(spectrum is M2 Iab)
- **Ib** (less luminous supergiants), Example:
- **II** bright giants, Example: β Scuti (HD 173764) (spectrum is G5 II)
- **III** normal giants, Example: ρ Persei (spectrum is O7.5 III(n) ((f)))
- **IV** subgiants, Example: ε Reticuli (spectrum is K1–2 IV)
- **V** main-sequence stars (dwarfs), Example: AD Leonis (spectrum M3.5e V)
- **VI** subdwarfs, Example: SSSPM J1930-4311 (spectrum sdM7)
- **VII** (uncommon) white dwarfs. White dwarfs are represented with a prescript wD or WD

Spectral types

Class	Conventional color	Actual apparent color
O	blue	blue
B	blue white	deep blue white
A	white	blue white
F	yellow white	white
G	yellow	yellowish white
K	orange	pale yellow orange
M	red	light orange red
L	red brown	scarlet
T	brown	magenta
Y	dark brown	black

Class **O** stars are very hot and extremely luminous, with most of their radiated output in the ultraviolet range. These are the rarest of all main-sequence stars. About 1 in 3,000,000 (0.00003%) of the main-sequence stars in the solar neighborhood are class **O** stars. Some of the lie within this spectral class. Class O stars frequently have complicated surroundings which make measurement of their spectra difficult.

Class **B** stars are very luminous and blue. Their spectra have neutral helium, which are most prominent at the B2 subclass, and moderate hydrogen lines. Ionized metal lines include Mg II and Si II. As O and B stars are so powerful, they only live for a relatively short time, and thus they do not stray far from the area in which they were formed.

These stars tend to be found in their originating OB associations, which are associated with giant molecular clouds. The Orion OB1 association occupies a large portion of a spiral arm of our galaxy and contains many of the brighter stars of the constellation Orion. About 1 in 800 (0.125%) of the main-sequence stars in the solar neighborhood are class **B** stars.

Class **A** stars are among the more common naked eye stars, and are white or bluish-white. They have strong hydrogen lines, at a maximum by A0, and also lines of ionized metals (Fe II, Mg II, Si II) at a maximum at A5. The presence of Ca II lines is notably strengthening by this point. About 1 in 160 (0.625%) of the main-sequence stars in the solar neighborhood are class A stars.

Class **F** stars have strengthening *H* and *K* lines of Ca II. Neutral metals (Fe I, Cr I) beginning to gain on ionized metal lines by late F. Their spectra are characterized by the weaker hydrogen lines and ionized metals. Their color is white. About 1 in 33 (3.03%) of the main-sequence stars in the solar neighborhood are class **F** stars.

Class **G** stars are probably the best known, if only for the reason that the Sun is of this class. They make up about 7.5%, nearly one in thirteen, of the main-sequence stars in the solar neighborhood.

Most notable are the *H* and *K* lines of Ca II, which are most prominent at G2. They have even weaker hydrogen lines than F, but along with the ionized metals, they have neutral metals. There is a prominent spike in the G band of CH molecules. G is host to the "Yellow Evolutionary Void". Supergiant stars often swing between O or B (blue) and K or M (red). While they do this, they do not stay for long in the yellow supergiant G classification as this is an extremely unstable place for a supergiant to be.

Class **K** stars are orangish stars that are slightly cooler than the Sun. They make up about 12%, nearly one in eight, of the main-sequence stars in the solar neighborhood. Some K stars are giants and supergiants, such as Arcturus, whereas orange dwarfs, like Alpha

Centauri B, are main-sequence stars.

They have extremely weak hydrogen lines, if they are present at all, and mostly neutral metals (Mn I, Fe I, Si I). By late K, molecular bands of titanium oxide become present. There is a suggestion that K Spectrum stars may potentially increase the chances of life developing on orbiting planets that are within the habitable zone.

Class **M** stars are by far the most common. About 76% of the main-sequence stars in the Solar neighborhood are class **M** stars. However, because main-sequence stars of spectral class M have such low luminosities, none are bright enough to be visible to see with the unaided eye. The brightest known M-class main-sequence star is M0V Lacaille 8760 at magnitude 6.6 (the fractionally brighter Groombridge 1618 was once considered to be class M0 but is now considered to be as K5) and it is extremely unlikely that any brighter examples will be found.

Although most class **M** stars are red dwarfs, the class also hosts most giants and some supergiants such as VY Canis Majoris, Antares and Betelgeuse, as well as Mira variables. Furthermore, the late-M group holds hotter brown dwarfs that are above the L spectrum. This is usually in the range of M6.5 to M9.5. The spectrum of a class M star shows lines belonging to oxide molecules, TiO in particular, in the visible and all neutral metals, but absorption lines of hydrogen are usually absent. TiO bands can be strong in class M stars, usually dominating their visible spectrum by about M5. Vanadium monoxide bands become present by late M.

VY Canis Majoris is a class M hypergiant. This has at times been reported as the largest known star, but its precise size is debated due to uncertainties over its distance, luminosity, and temperature. Artist's impression.

Andromeda

The Princess

Andromeda was the princess of Ethiopia, daughter of Cepheus and Cassiopeia. Cassiopeia was a boastful woman, and foolishly bragged that she was more beautiful than Hera, the queen of the gods, and the Nereids. In order to avenge the insult to his nymphs, Poseidon sent a sea monster to ravage the Ethiopian coast. (Some accounts state that the constellation Cetus represents the sea monster, but a more common view of Cetus is that he is a peaceful whale.)

Perseus tells Andromeda's parents that he'll kill the monster if they agree to give him their daughter's hand in marriage. They of course give him their consent, and Perseus kills the monster. (His exact method of doing so varies in different versions of the myth. Ovid has Perseus stab the monster to death after a drawn-out, bloody battle, while other versions have the hero simply hold up the head of Medusa, turning the monster to stone.) Andromeda is freed, and the two joyously marry.

The asterism consists of the brightest star, *Alpheratz* (or *Sirrah*) denoting Andromeda's head, and the rest of the principal stars marking other parts of the young woman's body. But I like to think that the other stars in fact trace Andromeda's flowing hair.

The Bayer stars are not very bright, as the constellation generally ranges from third and fourth magnitude stars.

There are a number of fine binaries and several variables, and some very nice deep sky objects, including perhaps the most famous spiral galaxy of all.

Double stars:

δ Delta (31) Andromedae: 3.27 magnitude; spectroscopic binary; suspected variable. A distance of approximately 105.5 light-years (32.3 parsecs) from the Earth.

η Eta Andromedae is a spectroscopic binary star consisting of two G-type subgiant or giant stars orbiting each other with an overall apparent visual magnitude of approximately 4.4.

γ Gamma^1 and γ^2 gamma^2 Andromedae form a noted binary with color contrast, gold and blue.

> The binary is a multiple system with a 2.5-5 magnitude.
>
> BC (the primary of which is gamma^2) form a very close binary with an orbit of 61 years: 5.5-6.3 magnitude.

ι Iota Andromedae has an apparent magnitude of 4.29. Iota Andromedae is a B-type main sequence star with a stellar classification of B8 V. It is among the least variable stars observed during the Hipparcos mission. It has been known by the name of **Keff al Salsalat**.

κ Kappa Andromedae is a wide and rather faint binary. Its apparent visual magnitude is 4.1-11. It is luminous enough to be visible from the suburbs and from urban outskirts, but not from brightly lit inner city regions.

λ Lambda Andromedae has a magnitude of 3.7-3.8.

μ Mu Andromedae has an apparent visual magnitude of 3.87. The star is situated about half way between the bright star Mirach to the southwest and the Andromeda Galaxy

(M31) to the northeast.

ν Nu Andromedae is a binary star with an apparent visual magnitude of 4.5. It is situated just over a degree to the west of this star is the Andromeda Galaxy. The primary component is a B-type main sequence star with a stellar classification of B5 V. The fainter secondary has a classification of F8 V, which makes it an F-type main sequence star.

σ Sigma Andromedae has an apparent visual magnitude of 4.5.

Tau Andromedae: 5, 10; 329°, 52.5".

π Pi Andromedae is faint and wide binary system with an apparent magnitude of +4.34. It is a spectroscopic binary with a companion star of magnitude 8.6.

φ Phi Andromedae has a 370 year orbit and has a magnitude range of 4.5-6.5.

ο Omicron Andromedae: A binary star, whose two components are both spectroscopic binaries themselves, making a four-star system. The system as a whole is classified as a blue-white B-type giant.

The separation of the two brightest components, **ο Andromedae A** and **ο Andromedae B**. A is a Gamma Cassiopeiae type variable star and the system's brightness varies from magnitude +3.58 to +3.78.

Omega Andromedae: 5, 12; 122°, 1.9".

υ Upsilon Andromedae: *A* binary star. The primary star (Upsilon Andromedae A) is an F-type main-sequence star that is somewhat younger than the Sun. The second star (Upsilon Andromedae B) is a red dwarf in a wide orbit around the primary.

ζ Zeta Andromedae has an apparent magnitude of 3.92 to 4.14.

51 Andromedae: The 5th brighest in the constellation Andromeda. It is occasionally called by the proper name **Nembus.** It is an orange K-type giant with an apparent magnitude of +3.59.

Variable stars:

α Alpha Andromedae is traditionally named **Alpheratz** (or Alpherat) and **Sirrah** (or Sirah), is the brightest star in the constellation of Andromeda. Located immediately northeast of the constellation of Pegasus, it is the northeastern star of the Great Square of Pegasus. It has an apparent magnitude of 2.07.

β Beta Andromedae: Known as **Mirach**, this Red Giant star has an average apparent visual magnitude of 2.05, which makes it the brightest star in the constellation. However, the luminosity varies slightly from magnitude *+2.01 to +2.10.* β Andromedae is located northeast of the Great Square of Pegasus and is theoretically visible to all observers north of 54° S.

M31 – Andromeda Galazy

The **Andromeda Galaxy, a**lso known as Messier 31, **M31**, or **NGC 224**, it is often referred to as the Great Andromeda Nebula. The Andromeda Galaxy is the nearest spiral galaxy to our Milky Way galaxy, but not the nearest galaxy overall. It is the largest galaxy of the Local Group, which also contains the Milky Way, the Triangulum Galaxy, and about 30 other smaller galaxies.

At 3.4, the apparent magnitude of Andromeda Galaxy is one of the brightest of that of any Messier object, making it visible to the naked eye on moonless nights even when viewed from areas with moderate light pollution. Although it appears more than six times as wide as the full Moon when photographed through a larger telescope, only the brighter central region is visible to the naked eye or when viewed using binoculars or a small telescope.

Other stars:

ε **Epsilon Andromedae** is a G-type giant star with an apparent visual magnitude of approximately 4.37.

NOTES

Antlia

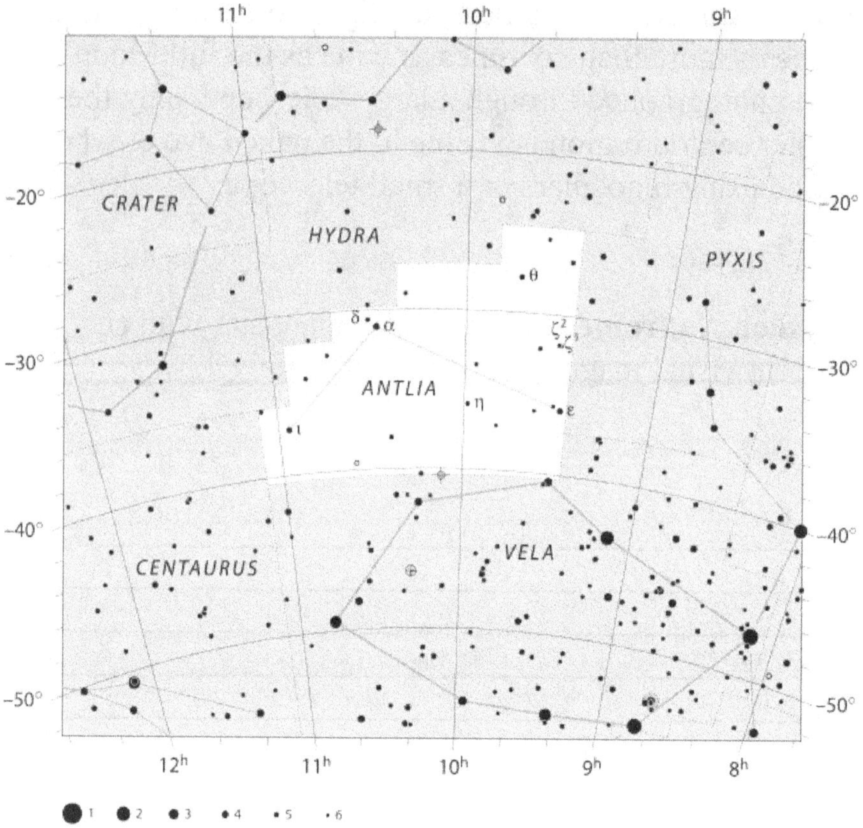

The Water Pump

Antlia is one of many constellations introduced by Nicolas Louis de Lacaille in the mid eighteenth century, designed to fill in the southern hemisphere.

Antlia is located in a rather bleak and lonely part of the southern hemisphere. It takes some imagination to find a "pump" here, not surprising perhaps, given the small selection of Bayer stars.

Double stars:

Zeta^{1A} and zeta^{1B} *Antliae* form a wide binary: 6.4, 7.2; PA 212°, separation 8".

Eta Antliae is an even wider binary with faint companion: 5.0 – 12 magnitude.

Variable stars:

U Antliae is an Lb irregular variable, ranging from 8.1 to 9.7 magnitude.

Other stars:

α Alpha Antliae is the brightest star in Antlia with a magnitude of 4.28 but it has not been given a proper name.

ε Epsilon Antliae has an apparent visual magnitude of this star is 4.5

Deep Sky Objects:

Antlia has many spiral galaxies, however they are all quite faint. One bright planetary nebula, *NGC 3132*, is sometimes assigned to Antila, however this object is now generally given to Vela.

Apus

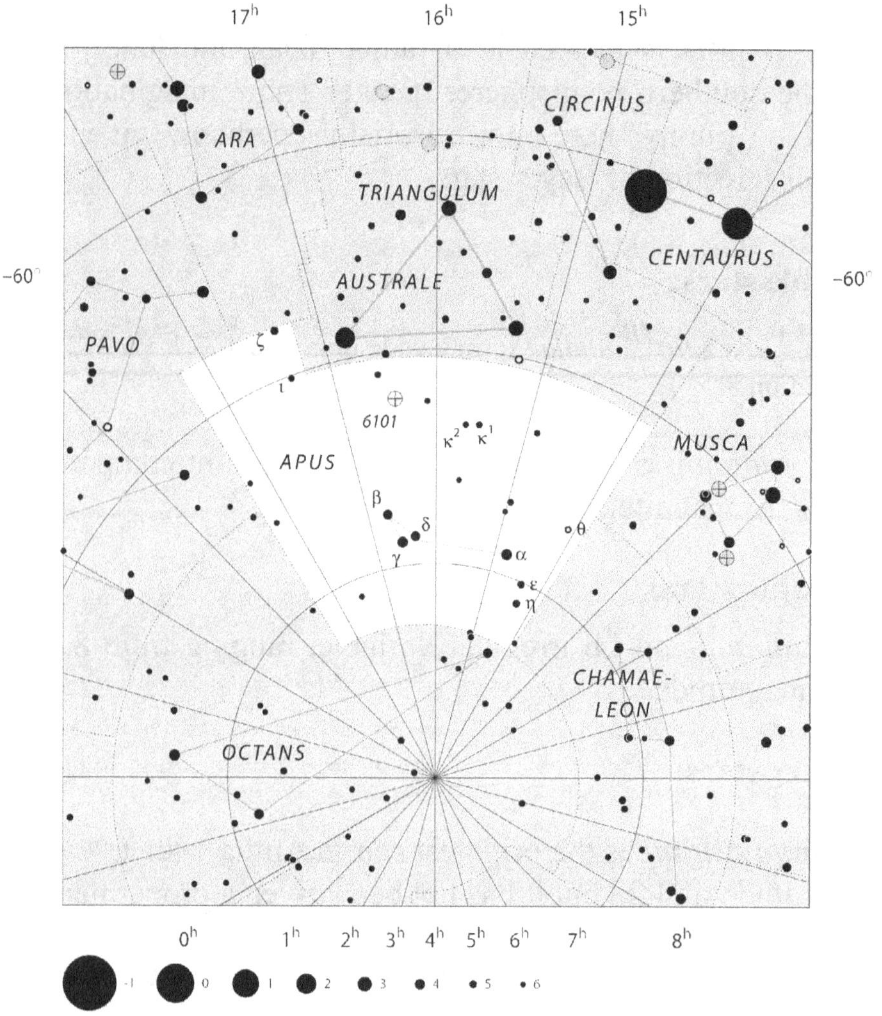

Bird of Paradise

Apus is one of those small and inconsequential constellations adapted from others in 1603 by Johann Bayer, designed to fill in the blanks in the Southern Hemisphere. Apart from several binaries and a faint globular cluster little else is found in this portion of the southern skies.

Apus, or Bird of Paradise, was known from sixteenth century voyagers, also being called "Apus Indica" or Bird of India. Some say it comes from the Greek apous, meaning without feet, as a reference to a Greek myth about the swallow, which was said to be legless.

From the paucity of interesting elements in this constellation, it might be argued that the name comes from the Greek apousia, which means "absence".

Double stars:

δ Delta^1 and *δ delta^2 Apodis* form a rather wide double of two orange giants: 4.7 – 5.27 magnitude.

Some observers report a slight colour difference, a reddish primary and an orange companion.

Variable stars:

Apus has three Bayer-star variables:

δ Delta^1 Apodis is an M-type red giant with a mean apparent magnitude of +4.68. It is classified as an irregular variable star and its brightness varies from magnitude +4.66 to +4.87. At an angular separation of 102.9 arcseconds is **δ² Apodis**, an orange K-type giant with an apparent magnitude of +5.27.

Theta Apodis is a semi-regular variable with a magnitude of 6.4-8.6.

k Kappa^1 Apodis is a gamma Cas variable: 5.43-5.61.

Other stars:

α Alpha Apodis has is the brightest star in the southern circumpolar constellation of Apus, with an apparent magnitude of approximately 3.83. This is a giant star with a stellar classification of K2.5III.

β Beta Apodis has an apparent visual magnitude of this star is 4.24. The spectrum of this star matches the characteristics of a K0 III

η Eta Apodis has an apparent visual magnitude of +4.9. The stellar classification of Eta Apodis shows this to be an Am star, which means the spectrum shows chemically peculiarities. In particular, it is an A2-type star showing an excess of the elements chromium and europium.

ε Epsilon Apodis has an apparent visual magnitude of 5.06, which is bright enough to be viewed from dark suburban skies. It has a blue-white glow that is a characteristic of B-type stars.

Based upon a stellar classification of B3 V, this is a massive, B-type main sequence star.

Epsilon Apodis is classified as a Gamma Cassiopeiae type variable star and its brightness varies between magnitudes 4.99 and 5.04.

γ Gamma Apodis has an apparent visual magnitude of 3.86. A stellar classification of G9 III identifies it as a giant star in the later stages of its evolution.

Deep Sky Objects:

NGC 6101 is a faint globular cluster, with stars no brighter than fourteenth magnitude. The cluster is found nearly 7° due north of gamma Apodis.

NOTES

Aquarius

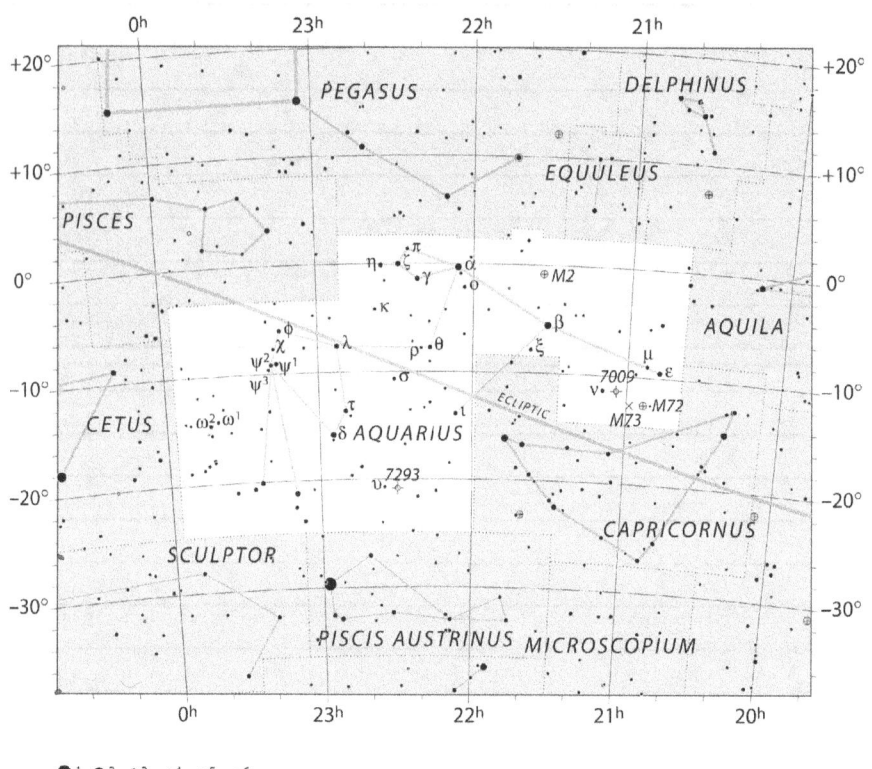

The

Water Carrier

The water carrier represented by the zodiacal constellation Aquarius is Ganymede, a beautiful Phrygian youth. Ganymede was the son of Tros, king of Troy (according to Lucian, he was also son of Dardanus). While tending his father's flocks on Mount Ida, Ganymede was spotted by Zeus. The king of gods became enamored of the boy and flew down to the mountain in the form of a large bird, whisking Ganymede away to the heavens. Ever since, the boy has served as cupbearer to the gods.

"Why is so much luck found in Aquarius", you may ask. When the sun entered Aquarius the new year was about to begin, Spring was on the horizon and the watery season would assure abundant crops. One can therefore appreciate the importance of the Water Bearer.

Incidentally, if the "Age of Aquarius" was celebrated in the 1960s, the real event is still some 600 years off: at that time Aquarius will contain the vernal equinox, marking the return of the Sun into the northern celestial hemisphere.

Single stars:

α Alpha Aquarii has the traditional name **Sadalmelik**, which is derived from Arabic for "Luck of the king". The apparent visual magnitude of 2.94 makes this the second-brightest star in Aquarius. The star glows with the yellow hue of a G-type star.

δ Delta Aquarii is the third-brightest star in the constellation Aquarius. It has the traditional name **Skat**, which has also been used for Beta Pegasi. The apparent visual magnitude is 3.3. The spectrum of Skat matches a stellar classification of

A3 V, indicating this is an A-type main sequence star.

ε Epsilon Aquarii has an obscure traditional name, **Albali**, from the Arab☐☐☐☐☐☐ *albāli'* "the swallower" and is visible to the naked eye with an apparent visual magnitude of 3.77. Epsilon Aquarii is an A-type main sequence star with a stellar classification of A1 V.

γ Gamma Aquarii has the traditional name **Sadachbia**, from an Arabic expression☐☐☐☐☐☐☐ ☐☐☐ *sa'd al-'axbiyah* "luck of the homes (tents)" in Hindu system it is also called Satabhishaj (a hundred physicians) in devnagari, sadhayam in tamil. This star has an apparent visual magnitude of 3.849, making it one of the brighter members of the constellation. Gamma Aquarii is an A-type main sequence star with a stellar classification of A0 V.

η Eta Aquarii has an apparent visual magnitude of 4.04. Eta Aquarii is near the radiant of a meteor shower named after it. It has a stellar classification of B9IV-Vn,

θ Theta Aquarii has the traditional name **Ancha**; Medieval Latin for "the haunch". Since it is near the ecliptic it can be occulted by the Moon, or very rarely by planets. Ancha belongs to the spectral class G8 with a luminosity class of III–IV suggesting that, at an age of 437 million years, this star is part way between the subgiant and giant stages of its evolution.

Ψ² Psi² Aquarii has an apparent visual magnitude of 4.4. This is a B-type main sequence star with a stellar classification of B5 Vn.

Double stars:

β Beta Aquarii has the traditional name **Sadalsuud**, from an Arabic expression ☐☐☐☐☐☐ ☐☐☐ sa'd al-su'ūd, the "luck of lucks".r identic with R.H. Allen), meaning the brightest of luck of lucks. Sadalsuud is the brightest star in Aquarius with an apparent magnitude of 2.87 and a stellar classification of G0 Ib.

ζ Zeta Aquarii has the traditional name Sadaltager (or Altager), from the Arabic☐☐☐☐☐☐ ☐☐☐ sa'd al-tājir "luck of the merchant". ζ^2 *Zeta² Aquarii* and ζ^1 *zeta¹ Aquarii* form a binary of two equal white stars with an orbit of 760 years. ζ^2 *Zeta² Aquarii* is the primary: With a magnitude ranging from 4.4 to 4.6.

Struve 2944 is a nice triple system, with all three in a neat line. The binary is 2° due east of kappa Aquarii. *Struve 2988* is a very attractive pair of equal stars: 7.2 and 7.2 magnitudes. The binary is 3° SW of psi^1 Aquarii.

Triple stars:

ψ¹ Psi¹ Aquarii also known as **91 Aquarii** has an apparent visual magnitude of 4.248. An extrasolar planet is known to orbit the main star.

Component	Apparent magnitude (V)	Spectral type
A	4.22	K0 III
B	9.62	K3 V
C	10.10	

*ω² **Omega² Aquarii** has* an apparent visual magnitude of 4.49. The primary component of this system is a massive, B-type main sequence star with a stellar classification of B9 V. There is a close orbiting stellar companion of unknown type, with a third component that is a K-type main sequence star with a visual magnitude of 9.5.

Binary stars:

*λ **Lambda Aquarii*** has the obscure traditional names **Hydor** and **Ekkhysis**, from the ancient Greek ὕδωρ "water" and ἔκχυσις "outpouring". The apparent visual magnitude of this star is 3.722. Lambda Aquarii is a red giant star with a stellar classification of M2.5 III.

Variable stars:

The most remarkable variable in the constellation is *R Aquarii*, usually listed as a "Mira variable". Yet this red giant isn't your normal long-period variable; it is a 'symbiotic star', resembling Z Andromedae.

> "Z Andromedae" stars are those which show two separate spectra, indicating two quite different temperatures, one cool, the other very hot. This phenomenon is caused by a very close binary system, which the larger star the cooler one, the small star(perhaps a white dwarf) the hot one.
>
> And in fact, R Aquarii has a small blue companion, which is encircled by a gas cloud. When this small star eclipses the giant, the visual magnitude of the primary drops several degrees.
>
> The star has a period of 386.96 days and a range from

5.8 to 12.4; the best time to view this star after the year 2000 is in 2005, in the first week of September.

Deep Sky Objects:

M2 (NGC 7089) is a globular cluster, compact and bright, about 50,000 light years away. The cluster is 5° N of *beta Aquarii*.

M72 (NGC 6981) is also a globular cluster, about 3° WSW of the Saturn Nebula (see below). It is one of Messier's least attractive objects.

M73 (NGC 6994) is another uninteresting Messier, a 'cluster' comprised of four unrelated stars about 1.5° east of M 72.

NGC 7009, "Saturn Nebula" is a planetary nebula quite spectacular in large instruments. It has 'rays' which extend from both sides of the main disc. The nebula is 1° west of *nu Aqr*. Burnham (p. 190) has a location chart.

NGC 7293, "Helix Nebula" (or the "Helical Nebula"), is another planetary nebula, given its name apparently because it is said to resemble the DNA double helix. It really is a ring nebula, only much larger and fainter than the more notable Ring Nebula in Lyra. The nebula is 1.5° W of *upsilon Aquarii*, or 21° due south of *zeta Aquarii*.

NOTES

_____
_____
_____
_____

Aquila

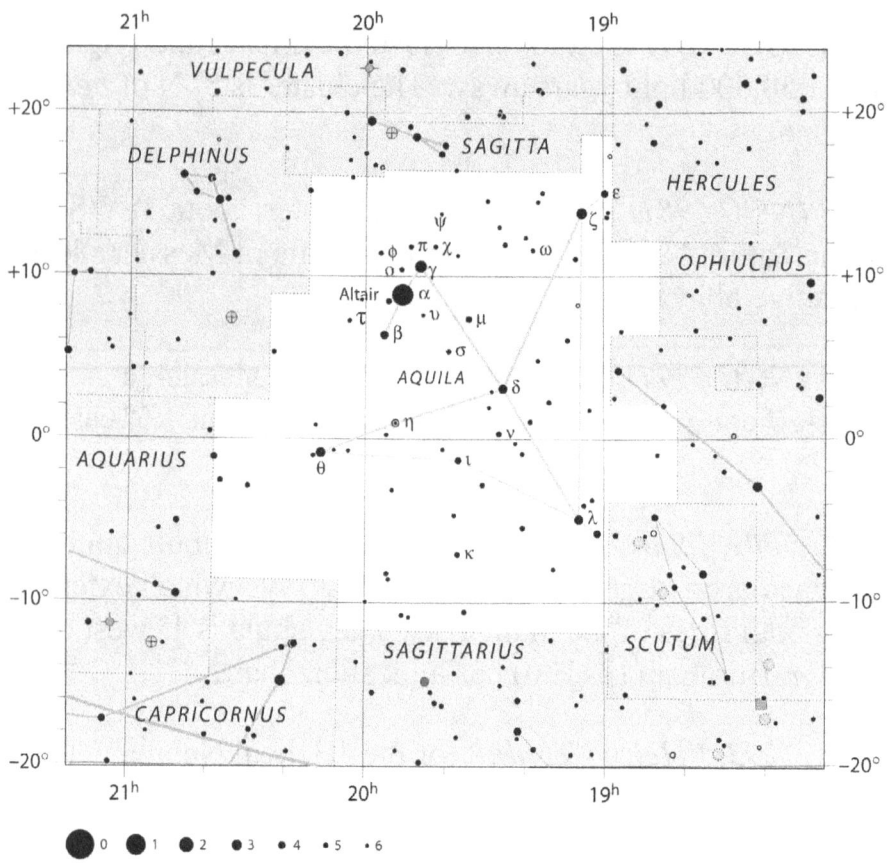

The Eagle

Aquila, The Eagle, is another ancient constellation whose history is linked to the Greek gods. The most often-told story is that of Hebe, daughter of Zeus and Hera, who married Heracles.

Before her marriage Hebe was the goddess of youth and she appeared in ceremonies as the official cup bearer. (That's to say, in all religious functions, she was responsible for pouring the wine.) She gave up the post after her marriage (although some accounts say that in one ceremony she indelicately exposed herself and was promptly sacked). In any case, the position was open and Zeus sought a suitable replacement. Ganymede was flown away by Zeus in the form of a giant Eagle.

The constellation is nearly out of sight from North America and Europe, as its stars extend from -46° to -60°. (In fact it goes much further south, however except for a faint globular cluster there's nothing of interest south of delta Ara.)

And yet this rather obscure constellation has a number of rather interesting deep sky objects.

NOTES

Ara

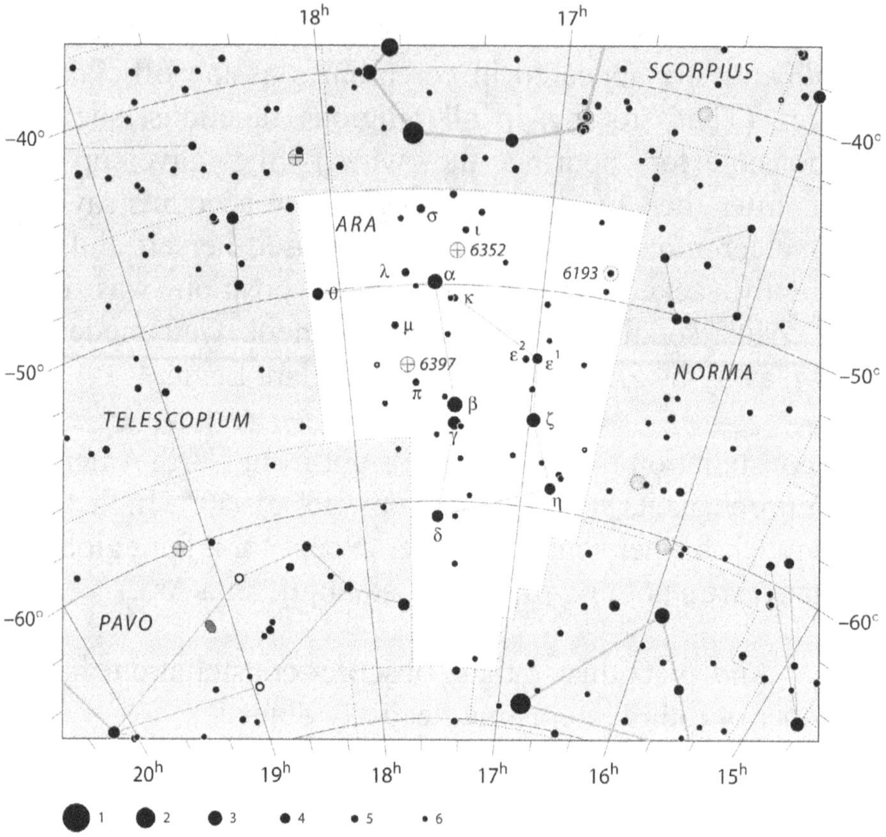

The Altar

Ara (The Altar) commemorates the altar on which sacrifices were made to the gods, in both Greek and Roman times. The Romans called it Ara Centauri, considering it to represent the altar Centaurus used (perhaps to sacrifice Lupus, the Wolf).

The constellation is nearly out of sight from North America and Europe, as its stars extend from -46° to -60°. (In fact it goes much further south, however except for a faint globular cluster there's nothing of interest south of delta Ara.)

And yet this rather obscure constellation has a number of rather interesting deep sky objects.

Double stars:

> *Gamma Arae* is a fixed binary with rather faint companion: 3.5, 10; PA 328, separation 18". A second companion is twelfth magnitude: PA 66°, separation 41.6". (Burnham questions whether these form a true binary.)
>
> *h4876* is a pleasant multiple in the star cluster NGC 6193 (see below): 6.6, 8.5; 14°, 1.6"; there is a seventh magnitude companion at PA 266° and 9.6".

h4866, better known as *R Arae* (see below): 6.0, 8.5; PA 123°, 3.6".

Variable stars:

The only variable of any interest in Ara is R Arae, an eclipsing binary which changes from 6.0 to 7.0 every 4.4 days.

Deep Sky Objects:

While there are no Messier objects, several clusters are of some interest.

> *NGC 6193* is a very large open cluster of about thirty stars, in Ara, located eight degrees west of alpha Arae and one degree north (or about seven degrees SSW of zeta Scorpii).

NGC 6397 is a bright globular cluster 2.5° NE of beta Arae. It is only visible to those living in latitudes south of 30° North (which means Florida in the US and none of Europe). This cluster is three° NE of beta Arae, or forty arc minutes east of epsilon Arae.

NOTES

Aries

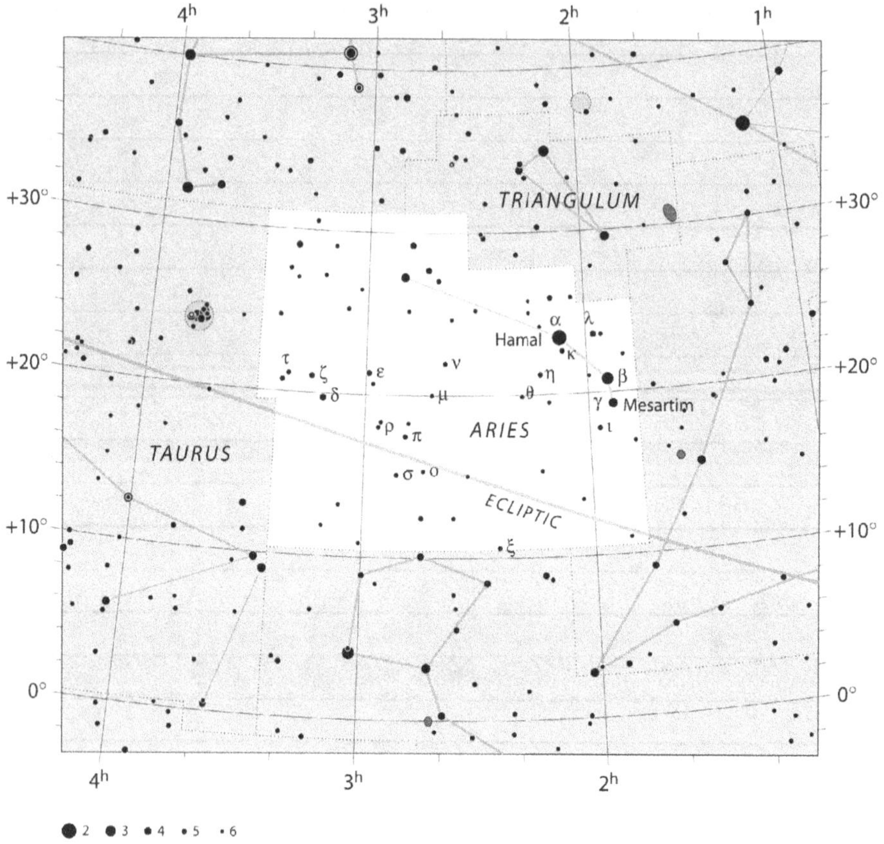

The Ram

Aries is a zodiacal constellation representing the ram of the Golden Fleece sought by Jason and the Argonauts. The ram had originally been presented to Nephele by Mercury when her husband took a new wife, Ino, who persecuted Nephele's children. To keep them safe, Nephele sent Phrixus and Helle away on the back of the magical ram, who flew away to the east. Helle fell off into the Hellespont (now the Dardanelles) between the Aegean Sea and the Sea of Marmara, but Phrixus safely made it to Colchis on the eastern shore of the Black Sea. Phrixus sacrificed the ram and presented the Golden Fleece to the king, Aeetes.

Aries' stars are rather faint except for alpha and beta, which are only second magnitude stars.

Double stars:

Gamma Arietis (Marsartim) is a well-known binary of similar stars: 4.8, 4.8; PA 360°, separation 7.8".

Lambda Arietis is a wide binary: 4.9, 7.7; PA 46°, separation 37.4".

Epsilon Arietis is a closer binary of nearly equal stars: 5.2, 5.5; PA 203°, separation 1.4".

30 Ari is a fixed binary with wide component: 6.6, 7.4; PA 274° and separation 38.6".

33 Ari is also fixed, with a faint component: 5.5, 8.4; PA 360°, separation 28.6".

Variable stars:

Gamma^2 Arietis is an alpha CV type variable: 4.62-4.66

with a period of 2.6 days.

SX Arietis (56 Ari) is the prototype of a special class of rotating variables, similar to alpha CV variables. SX Ari varies from 5.67 to 5.81 every 17h28m.

Deep Sky Objects:

NGC 772 is a strangely shaped diffuse galaxy with a spiral arm on the northwest.

It's found about one degree ESE of *gamma Ari*.

NOTES

Auriga

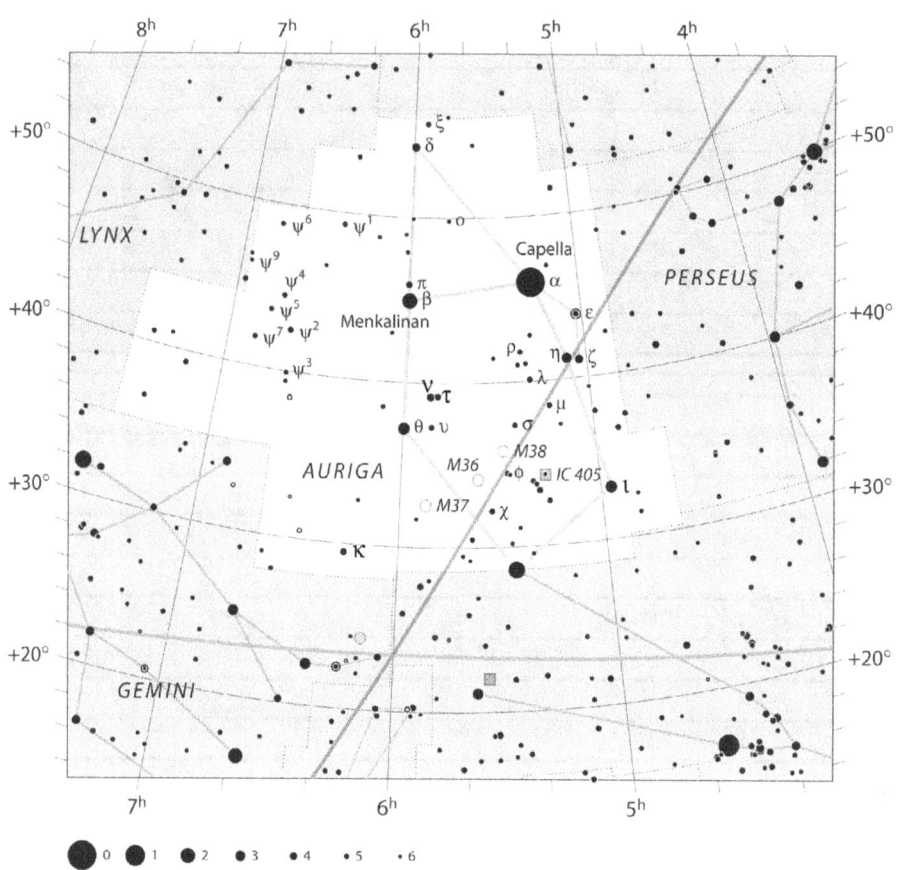

The Charioteer

Auriga is an ancient Northern Hemisphere constellation featuring one of the brightest stars in the heavens: Capella. Auriga is usually pictured as a charioteer; the youth Auriga wields a whip in one hand and holds a goat (Capella) and her two kids in the other.

To find Auriga, first locate Orion. Taurus is to the right (west) and just above these two, much higher in the sky, you will see Capella. While this star marks roughly the mid-point of the constellation, north to south, most of the more interesting aspects of the constellation are found to the south of the star, all the way down to El Nath, the second brightest star (gamma Aurigae) which is actually shared with Taurus, and also known as beta Tauri.

Auriga's stars are fairly bright; five are second magnitude or brighter. Alpha Aurigae (Capella) is the sixth brightness star, at a visual magnitude of 0.08. The star is 43.5 light years away, and is about ten times the size of our Sun.

Capella's visual magnitude is really the combined brightnesses of the primary star and a close companion, that revolves every 104 days. There is another companion, much fainter: a red dwarf which is itself a close binary.

Binary stars:

Zeta Aurigae is an eclipsing binary; an orange giant primary with a blue companion that orbits every 972 days (2.7 years).

Theta Aurigae is visible in large scopes: 2.6, 7.1; PA 300° and separation 3.6".

Omega Aurigae: 5.0, 8.0; PA 360°, separation 5.4".

14 Aurigae is a multiple double, visible in larger scopes.

> The primary is 5.1, with three companions: B (11.1, 352°, 11"), C (7.4, 225°, 15") and D (10.4, 356°, 7.7").

Variable stars:

There are a half-dozen variable stars in this constellation which are visible in small scopes, most of them of very small variance.

> *Epsilon Aurigae* is an unusual variable which normally maintains a visual magnitude of 2.92 but every 9892 days (27 years) dips down to 3.83.
>
>> The next scheduled dip is in the late summer of 2010. The eclipse phase lasts about a year.
>
> *R Aurigae* is the only Mira-type variable of interest. Normally a rather faint 6.7, every 457.5 days it takes a nose-dive to 13.9. The best time to view this feature is in late November of 2001, when it should be near the transit.

Deep Sky Objects:

Auriga has three Messier objects: M36, M37, and M38. A telescope is preferred but you can at least locate these objects with binoculars.

> *M36* is a rather faint cluster of about 50 to 60 stars, in a very compact area. A large scope is necessary to resolve the individual stars.
>
>> To find M36, move west just across an imaginary

line from El Nath to theta Aurigae.

M37 is the most spectacular of the three Messiers, and also the most easily found, as it lies midway between El Nath and theta Aurigae.

> This last star is to the east of El Nath and north, about half way up to Capella. Now slightly to the east of an imaginary line between these two stars, and half way along that line, is M37, a rich star cluster of perhaps 150 stars.
>
> Binoculars will only show a fuzzy mess; you really need a scope for this one. A medium sized scope should reveal at least twelve red giants, with the brightest one found at the centre of the cluster. Some observers find this star more orange than red. What do you think? In any case, it's a sight worth seeing. The cluster is considered to be about 200 million years old.

M38 is in the same field, just to the NW of M36. Some observers have described this cluster of about a hundred stars as having a cross-shape.

NOTES

_____
_____
_____
_____
_____
_____
_____

Bootes

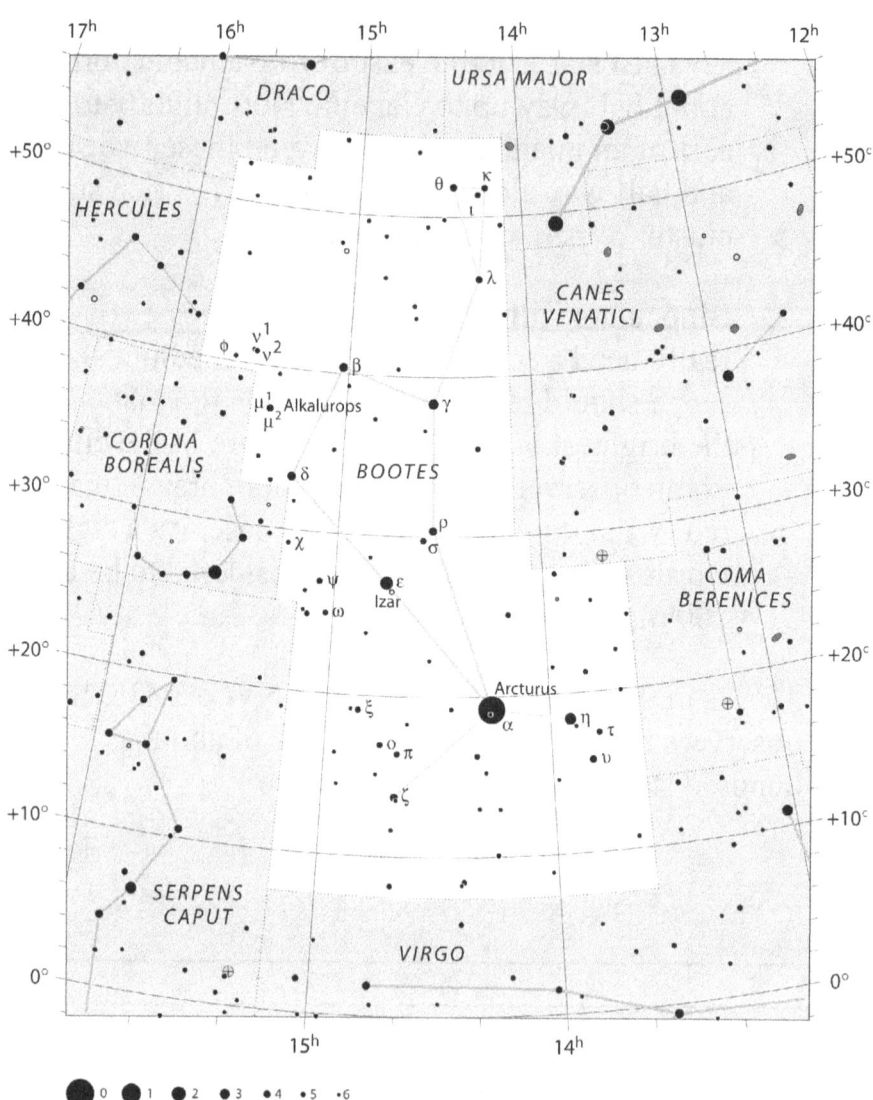

The Herdsman

Boötes may be a hunter, on the tracks of the Great Bear, accompanied by his two dogs Asterion and Chara (the "Canes Venatici"). And yet the constellation was once known as Arctophylax which means the protector of the Bear. Perhaps it was the Romans who changed his role, for they called him Venator Ursae: the Bear Hunter.

Nowadays Boötes is generally considered to be a Herdsman (as in French: Le Bouvier), as he eternally shepherds the stars around the North Pole.

The constellation was known in antiquity, with the first recorded appearance being in Homer's Odyssey. In Book V Odysseus sails his ship by the stars, using the Pleiades, the Bear, and Boötes ("which set late") to reach his destination.

The constellation is quite compact, squeezed between Canes Venatici and Hercules, with Virgo to the south. The northern border touches both Ursa Major and Draco. There is a full complement of Bayer stars.

To find the major star, Arcturus, follow the sweep of the Big Dipper's handle. These stars lead to the brightest star in the northern hemisphere, and the fourth brightest in all of the heavens: *alpha Boötis*, better known as *Arcturus*.

Arcturus means "Guardian of the Bear". This orange-red giant is about 20-25 times the size of the Sun, with about the same mass. In fact the Sun will probably take the same path, eventually ballooning to the same size in another five billion years.

The star has an unusually high proper motion (2.281") and a space velocity of 118 km/sec. It is 35.4 light years away.

Some constellations are known for their deep sky objects; others for an interesting variable or perhaps an attractive binary. Boötes has few deep sky objects of any interest. Nor are its variables particularly noteworthy. However the constellation does have one of the finest collection of double stars, some of which are described below.

Double stars:

Zeta Boötis is a very fast binary with a highly eccentric orbit of 123.4 years. The companion is currently fairly close (0.9") at PA 301°.

Epsilon Boötis was one of Struve's favourite double star systems: 2.5, 4.9; a bright yellow primary with a blue-green companion. The orbit is so large it may as well be considered fixed: PA 339°, separation 2.8".

Kappa Boötis is a gorgeous double with colour contrast; the primary is yellow and the companion a deep blue. The binary is fixed at PA 235°, separation 13.4".

In the same field is *iota Boötis*: 4.9, 7.5; PA 33°, separation 38.5"

Mu Boötis is a triple system. AB are fixed: 4.3, 7.0; PA 171°, separation 108". The component B has a close companion C (magnitude 7.6) which is a rapid binary, with an orbit of 246.1 years. The 2000 values are PA 7°, separation 2.1".

Pi Boötis is a pleasant binary of two blue-white stars (4.9, 5.8; PA 108°, separation 5.6").

Xi Boötis is a rapid binary (4.7, 7.0) with orbit of 151 years. The primary is yellow, and the companion a white-pink. Currently the companion is found at PA 321° and 6.8" separation.

Struve 1785 is another attractive rapid binary, with orbit of 155 years. The companion can now be found at PA 173°, 3.3" separation.

Struve 1909 (44 Boo) is yet another rapid double, with an orbit of 225 years. In the year 2000 the companion will be at its widest separation for the next fifty years. Presently it's located at 54°, 2.2" separation.

Variable stars:

The constellation contains three delta Scuti type variables: *gamma, iota,* and *kappa^2*.

> *Delta Scuti* variables are fairly young stars which pulsate, creating a slight variation in visual magnitude, less than 0.5 magnitude and often considerably less than this, with a short period (from 30 minutes to about eight hours). Thus the stars, while quite numerous, are not of the kind which amateur enthusiasts tend to study.

R Boötis is a long-period variable with a range from 6.2 to 13.1 every 223.4 days. In 2000 the maximum is scheduled to appear in early April.

Deep Sky Objects:

There are no Messier objects in this constellation, but for the inveterate deep sky observer there are still a few nice galaxies. Below are two of the better deep sky objects.

> *NGC 5248* is a very compact spiral galaxy in the southwestern corner of the constellation, ten degrees south of Arcturus and one and a half degrees west. (Two degrees north is the rapid binary BU 612 (6.3, 6.3) with period 22.4 years.)

NGC 5466 is a large but quite dim globular cluster. It's found nine degrees north of Arcturus and one and a half degrees west. (The Messier object M3 is five degrees due west in Canes Venatici.)

NOTES

Caelum

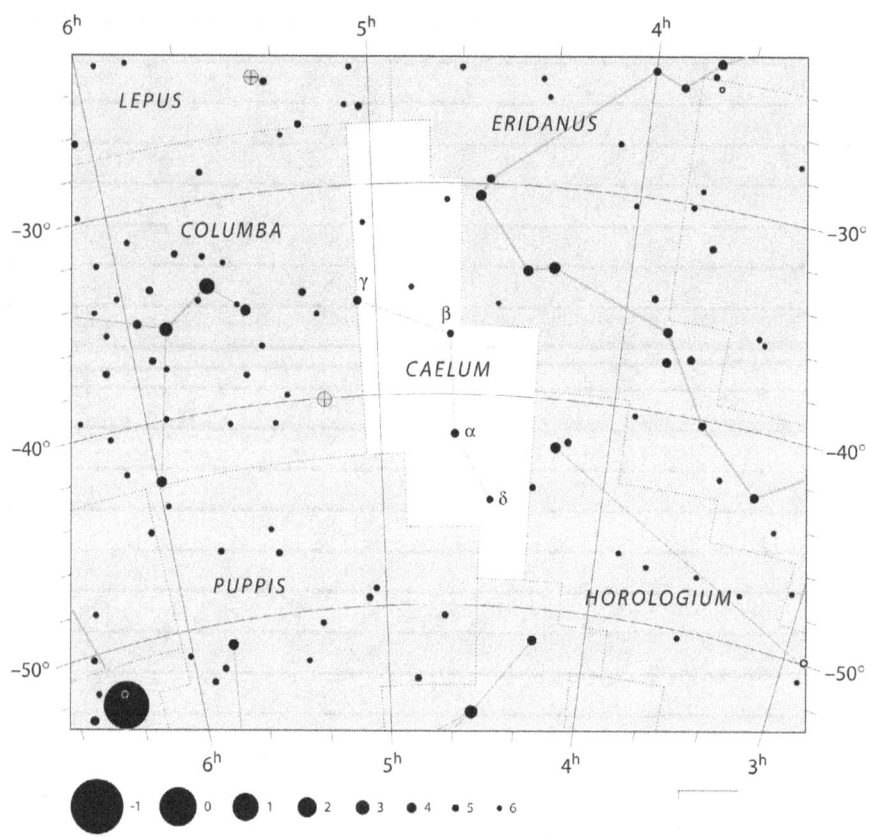

The Heavens

Like many southern hemisphere constellations, Caelum was introduced by Nicolas Louis de Lacaille in the mid-eighteenth century, designed to fill in the southern hemisphere.

The word is ambiguous; in Latin caelum means both "the heavens" and "burin", which is an instrument used for engraving on copper and fine metals. It is this instrument that Lacaille had in mind when he named the constellation.

In fact Lacaille drew two of these instruments in his original map, calling the constellation "Les Burins". Only one has remained.

Caelum is located between Eridanus and Columba in a particularly bleak part of the southern hemisphere. There are few Bayer stars here, and none brighter than fourth magnitude.

Double stars:

Alpha Caeli has an extremely faint companion: 4.5, 13; PA 121°, separation 6.6".

Gamma Caeli also has a rather faint companion: 4.6, 8.0; PA 308°, separation 2.9".

Variable stars:

R Caeli is a Mira-type long period variable, from 6.7 to 13.7 every 391 days. It's located one degree south of beta Caeli, and several arc minutes west.

X Caeli is a delta Scuti variable, ranging from 6.3 to 6.4

every 3h 14.7m.

Once labelled gamma^2 Caeli, the star now generally goes under its variable designation.

Deep Sky Objects:

Caelum has no deep sky objects (at least mentioned in Tirion's *Sky Atlas*). Burnham indicates one spiral galaxy, *NGC 1679*, which would be about two degrees south of zeta Caeli.

NOTES

Camelopardalis

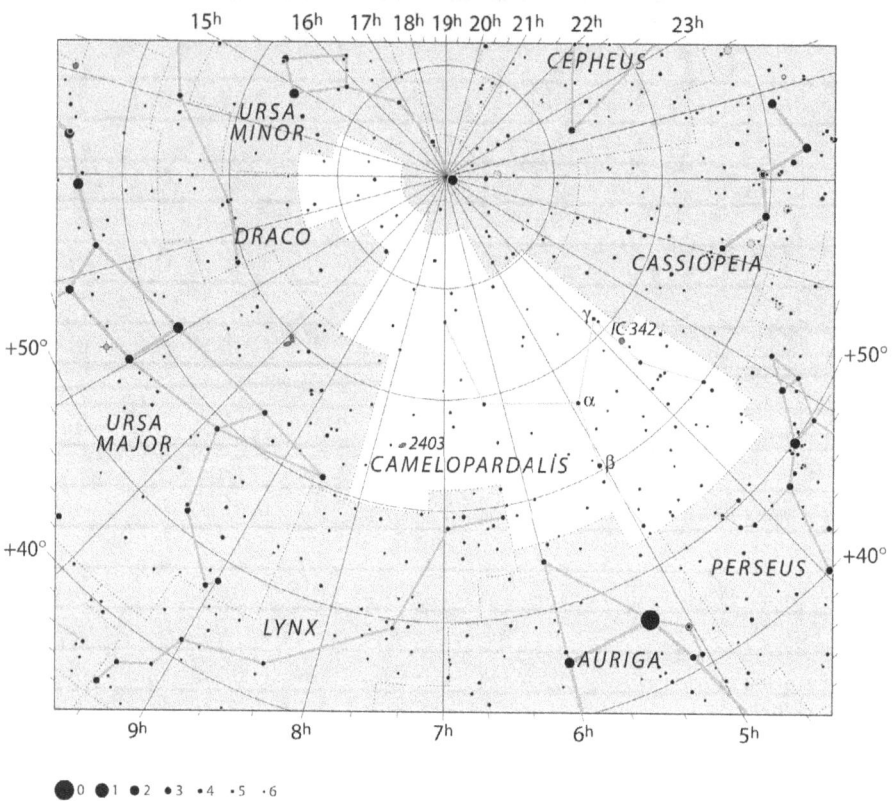

The Giraffe

If one were asked to name all the four-legged creatures found in the sky, the Ram and the Bull would come readily to mind, and the Bear and Dog (two of each actually: major and minor). A little more thought might produce the Hare (or Rabbit) and the Unicorn (however mythic it might be). Then some might recall that there is also a Fox and a Wolf. And yes, could there also be a Camel?

Not really. The Camel doesn't belong in our menagerie. Camelopardalis means Giraffe. It is also sometimes written Camelopardus, although the correct spelling is indeed CAMELOPARDALIS.

In the winter months the Giraffe appears upside down. You might want to study Camelopardalis in the summer, when it's right side up.

> The constellation was probably invented by Petrus Plancius (1552-1622), a Dutchman who made his name in cartography while working for the Dutch East India Company. His world maps of 1592 and 1594 became very popular, while his contribution to the heavenly maps was awarded in 1624 when Camelopardalis was included in Jakob Bartsch's book on the constellations. (Some historians believe Bartsch to have invented the constellation.)

While Camelopardalis sits between Cassiopeia and Ursa Major, the best way to begin studying it is to first find your bearings. With the naked eye locate *Capella (alpha Aurigae)*. If you aren't sure which of the bright stars is Capella, start from the Big Dipper. Now instead of moving north to the Pole Star, move across the top part of the dipper (a line from delta Ursae Majoris drawn through alpha Ursae

Majoris) and continue straight into the southern portion of the skies. As you approach the Milky Way, the very bright star you see is Capella.

Moving northwest from Capella you enter Perseus. Half way between Capella and Algenib (alpha Persei) and five degrees north of this last star, are the feet of the Giraffe. Roughly half way between Algenib and the North Pole is *gamma Camelopardalis*, the haunch of the giraffe.

Return to Capella; move west three degrees and north seven degrees. This is *7 Cam*, a binary (Struve 610) which serves as the giraffe's front foot.

Now that we've got his backside and front foot sorted out, let's move from *7 Cam* to the first bright star, about seven degrees north. This is *beta Cam*, also a binary (see below).

> *Beta Cam* is the brightest star in Camelopardalis, at 4.03 visual magnitude, a yellow supergiant roughly a hundred times the size of the Sun, and about 1700 light years away.

Further north another six degrees and you encounter *alpha Cam*, which is nearly as bright at 4.3. This is a blue supergiant 4000 light years distant, with a diameter about half that of *beta Cam*.

Northwest of *alpha Cam* is *gamma*, with a visual magnitude of only 4.63. This star is only twice the size of our Sun, and is about 180 light years away.

These are the only Bayer stars in the constellation. But that's not to say there aren't other stars of great interest.

NGC 1502 is found half way between *alpha* and *beta Cam*, and about 55 arc minutes west. The brightest star in this group is the primary of *Struve 485*, found at the centre. The binary's vital statistics are AB: 6.1, 6.2; PA 304 degrees, at separation 18.1".

At virtually the same location is a second binary, *Struve 484*, which is much fainter: AB: 9.0, 9.5; PA 132, 5.3".

Struve 1051 is a striking triple system of similar stars. AB: 6.5, 7.7; PA 284 degrees, separation 1.1"; C: 7.8, PA 82 degrees and separation 31.5".

This very nice system is found in an otherwise desolate region: 7h, 26m, 35s; +73 degrees, 4', 58". It's well worth the detour.

Struve 1694 is a wide pair of nearly equal stars (5.0, 5.5; PA 326 degrees, separation 21.6")

Beta Camelopardalis features a pale yellow primary and a very wide, much fainter, companion: 4.0, 9.0; PA 208 degrees, separation 80".

Component B has a closer companion, named "b", an 11-magnitude star at 14.8" and PA 168 degrees.

Variable stars:

R Cam is a Mira-type variable with a period of 270.22 days, rising from 14.4 visual magnitude only to about 7, which makes it a telescopic variable all throughout its cycle.

VZ Cam is a semi-regular with an average period of 23.7 days, varying from 4.80 to 5. This is a popular semi-regular for binoculars.

Deep Sky Objects:

Although there are no Messier objects in Camelopardalis, there are many galaxies and star clusters (most of which however are quite faint).

NGC 1502 is the finest star cluster, a small group of perhaps fifteen stars with the binaries Struve 484 and Struve 485 at its centre (see above).

Kemble's Cascade is a string of mostly eighth-magnitude stars (nicely seen in binoculars) which seem to "splash" into the cluster.

The asterism is named for Father Lucian Kemble, a Franciscan and avid Canadian amateur astronomer who first drew attention to it in the late 1970s. (We regretfully note that Father Luc died of heart failure in the early hours of the 21st of February 1999.)

NGC 2403 is a fine spiral galaxy about 10 million light years away. At ninth magnitude it's easily seen in medium sized telescopes, although greater detail is of course obtained in larger scopes.

NGC 2523 is an extremely faint barred spiral galaxy with very curious features. With a visual magnitude of 13, it is only accessible to larger telescopes.

NOTES

Cancer

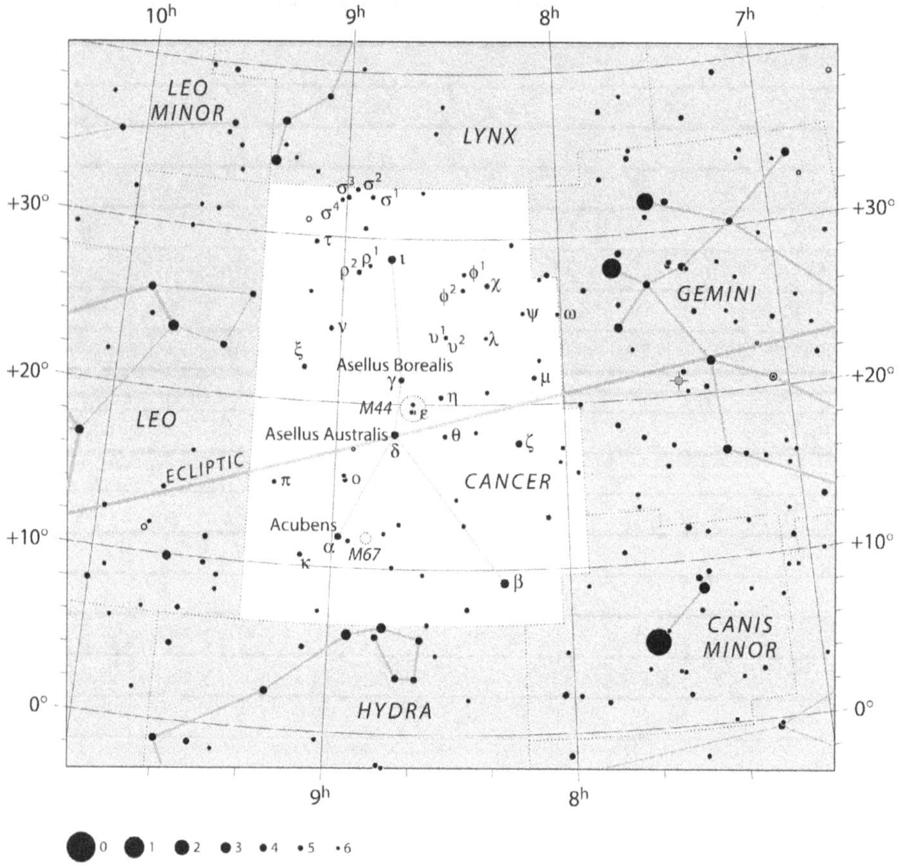

The Crab

𝕮ancer is a zodiacal constellation. As with many other constellations, its exact mythological origin is uncertain; however, the most widely accepted story is that Cancer was the crab sent to harass Heracles while he was on his second labor. As he battled the Lernaean Hydra, the ever-jealous Hera sent Cancer to nip at the hero's heels. The crab was eventually crushed beneath Heracles's feet, but Hera placed it in the heavens as a reward for its faithful service.

Cancer is a faint constellation located just east of Gemini and north of the head of the sprawling constellation Hydra. Its stars are generally four magnitude, beta Cancri being the brightest at 3.52.

While rather small, Cancer still has a number of fine objects, including a splendid star cluster and several visual binaries.

Double stars:

Zeta Cancri is a notable triple system comprised of a close binary with a period of 59.5 years and a more distant star, component C, with a much longer period of 1115 years. (These values are recently published revisions; formerly the two orbits were thought to be 59.7 and 1150 years.)

This distant companion also has its own binary star, which revolves about zeta^C every 17.6 years. It has never been seen, and its existence has only been discovered through a particular wobble of zeta^C. The unseen star is thought to be a white dwarf.

Phi Cancri is a binary of two identical white stars

(5.5m, 6m): the PA is 217° and separation is 5.1".

Iota Cancri is a wide binary (4.5, 6.5) with a striking colour contrast: yellow and blue. PA 307° separation 30.5".

Finally, for the perseverant, there are a number of binary systems visible in the *Beehive Cluster* (see below). We'll point out two of them, very close to each other.

> The brightest is *Struve 1254*. The primary is a bright 6.5m, with a 9.0m companion B at 54°, 20.5". Then there are two more components: C: 8.0, 342°, 63.2"; D: 9.0, 43°, 82.6".
>
> To find this group, first locate epsilon Cancri, which is near the centre of the Beehive Cluster and the brightest star in this cluster. Just to the northwest of this star, less than a minute's distance, you'll find this binary system.
>
> In the same field (slightly west and less than a minute south of *Struve 1254*) is the nice quadruple called *beta 584*, comprised of 7.0, 12.0, 7.0, and 6.5 visual magnitudes.
>
> AB is the most difficult to find, for the companion is a faint 12m star at 291° and separation of only 1.2". AC: 156°, 45"; AD: 241°, 93".
>
> To find this group, first locate epsilon Cancri, which is near the centre of the Beehive Cluster

and the brightest star in this cluster. Just to the northwest of this star, less than a minute's distance, you'll find this binary system.

In the same field (slightly west and less than a minute south of *Struve 1254*) is the nice quadruple called *beta 584*, comprised of 7.0, 12.0, 7.0, and 6.5 visual magnitudes.

> AB is the most difficult to find, for the companion is a faint 12m star at 291° and separation of only 1.2". AC: 156°, 45"; AD: 241°, 93".

Galileo was the first to study its stars with a telescope. He counted over forty members, putting to rest the idea of its nebulosity and introducing the idea of star clusters.

There are over three hundred stars in the Beehive (the Webb Society Handbook claims 2000). It has been estimated that over a hundred of its stars are brighter than our Sun, and in fact (as Burnham points out) if the Sun were a member of this group, it would be a very modest member indeed, at about 10.9 magnitude.

M67 (NGC 2682) sits about two degrees west of alpha Cancri and south of the Beehive about nine degrees.

> Visually unremarkable, yet this deep sky object is renowned for its venerable age: it is now believed that the cluster is approximately 10 billion years old. Its estimated distance is 2500 light years and there are about five hundred stars in the cluster,

tightly packed.

Being so old, many of its stars have nearly completed their life-cycle, having passed through the red giant stage and now having "jumped off" the main sequence and entering another phase.

Indeed, this is how the age of such clusters is determined.

It is assumed that all members of a star cluster evolved out of the same gas cloud at roughly the same time (give or take a few million years). These stars spend a given length of time on the main sequence, relative to their mass. For example, stars equal to one solar mass will spend about ten billion years on the main sequence. Since the stars of the Beehive Cluster are rather similar to the sun, its age has therefore been calculated to be at least 10 billion years.

NOTES

Canes Venatici

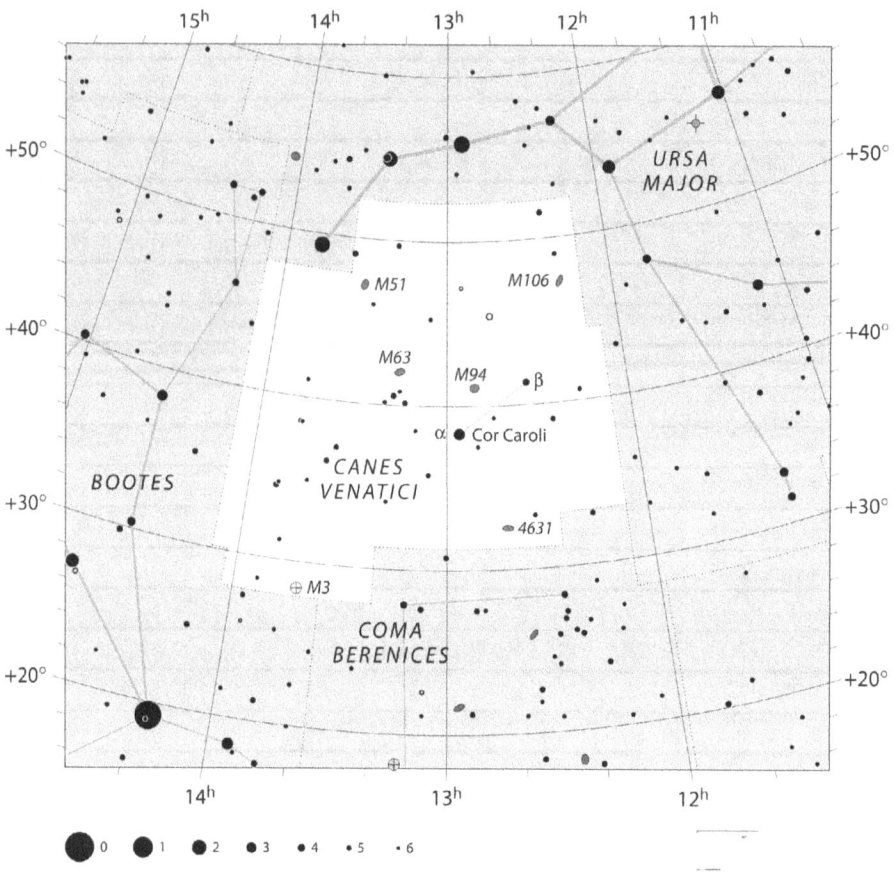

The Dogs

𝕮anes Venatici is one of those obscure constellations introduced by Johannes Hevelius in 1690. It represents the two dogs Asterion and Chara, both held on a leash by Bootes as they apparently chase the Great Bear around the North Pole.

With one exception, the constellation's stars are quite faint, fourth- and fifth-magnitude stars. There are only three Bayer stars, yet several notable binaries can be found, as well as a famous variable and a number of interesting deep sky objects as well.

Alpha Canum Venaticorum is popularly called Cor Caroli (*Heart of Charles*). Most sources give Edmund Halley the credit, naming it after King Charles II after the restoration of the monarchy in Britain in 1660. (Some say, however, that the reference was initially meant to commemorate Charles I, after his execution.)

The star has a visual magnitude of 2.9 (variable), a distance of 110 light years, and roughly the same size as our Sun. It is also a splendid double with, perhaps, a subtle colour contrast (discussed below).

Double stars:

Canes Venatici has two attractive binaries: *alpha CVn* and *25 CVn*.

Alpha^2 and *alpha^1 CVn* form a celebrated fixed double star system. Note that the primary is *alpha^2*, since it is slightly east of its companion.

While both stars are usually reported to be blue-white, some find them slightly different, perhaps soft blue and yellow, or two shades of white.

25 CVn (Struve 1768) is a visual binary with an elegant orbit of 240 years. Presently, the companion is at near maximum separation, with a PA of 100 degrees and separation 1.8".

Variable stars:

The constellation contains one of the more interesting semi-regular stars, Y CVn, called *La Superba* by its admirers. One look and you will understand why: the star is an unusually vivid red.

> *Y CVn* is classified as an SRb star. Such stars are known to have several periodic cycles, superimposed on each other. Basically, it changes in magnitude from 7.4 to 10.0 every 157 days. (However an update published in Budapest, in the *Information Bulletin of Variable Stars*, #2271, has reassessed the period at 251.8 days.)

Alpha^2 CVn is the prototype of a class of variables. Such stars usually have a spectrum from B9 to A5, are unusually abundant in particular heavy metals and deficient in common elements. Alpha^2 has an abundance of silicon, europium, and mercury, and oscillates in magnitude from 2.84 to 2.98 every 5.5 days. anging onto one of its extended arms.

> *M63* is shometimes called the Sunflower Galaxy, by its numerous arms, which Burnham describes as "reminiscent of showers of sparks thrown out by a

rotating fiery pinwheel". Fairly bright, at 8.1 magnitude, it has a very condensed centre. The galaxy is found five degrees north-northeast of Cor Caroli.

M94 is another spiral, seen practically face-on, and sometimes described as "comet-like". This is a very compact circular spiral and very bright (8.1 magnitude). To find it draw a line between Cor Caroli and beta CVn, and at the half-way point draw a perpendicular off to the northeast. About two degrees up this perpendicular is found M94.

M106 (NGC 4258) is another bright spiral. Burnham doesn't list this object as a Messier, but gives a fine photograph (p 375). The galaxy is six degrees north north-west of beta CVn.

Below are listed a selected number of galaxies considered the best of the non-Messiers.

NGC 4244: a large edge-on spiral, found eight degrees west of Cor Caroli.

NGC 4485 and *NGC 4490* are two splendid galaxies in the same field: 4485 is more compact (this one is sometimes called the Cocoon Galaxy), while 4490 is larger and brighter. Located less than one degree northwest of beta CVn.

NGC 4631: very large and bright, seen edge-on. Found in a rather barren field, six degrees south of Cor Caroli and two degrees west. In the same field are two more galaxies, NGC 4656 and 4657, just southwest of 4631.

Canis Major

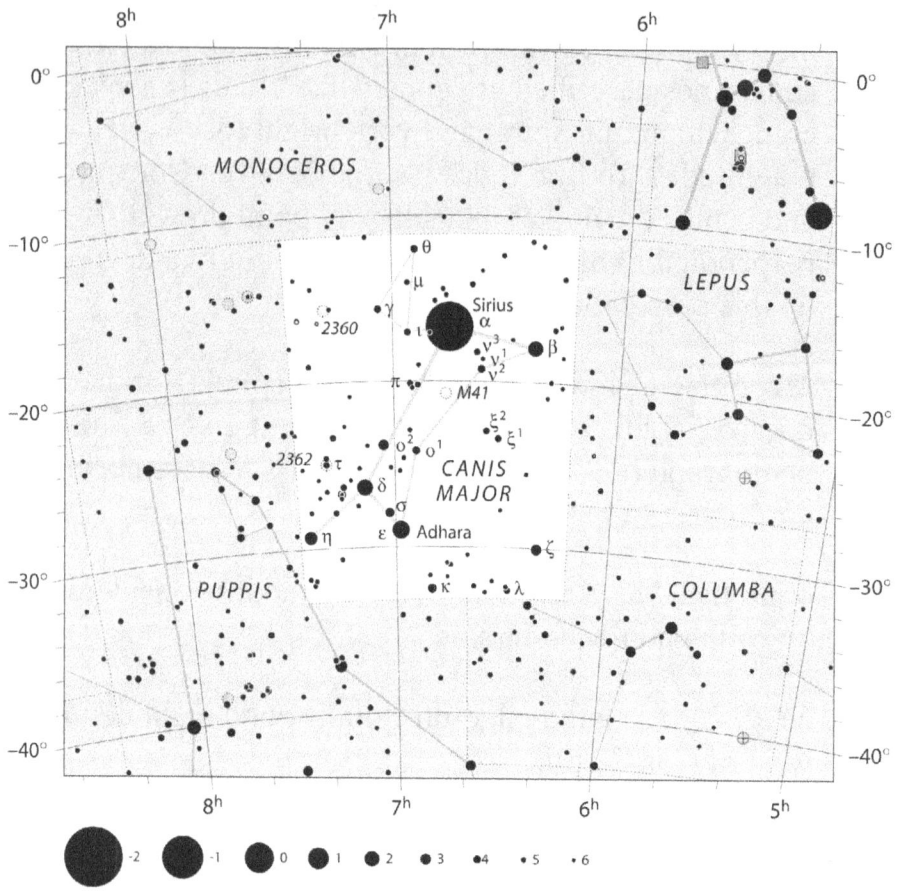

The Big Dog

Canis Major, the largest of Orion's two hunting dogs, might be chasing Lepus, the Rabbit, who is just in front of him. Or perhaps he is ready to help Orion battle the great bull.

The stories concerning Orion's dogs are not of mythic proportion, but the Greeks did have several interesting beliefs concerning Sirius, alpha Canis Majoris.

The Athenian New Year began with the appearance of Sirius. He was seen as two-headed, like the Roman God Janus: looking back at the past year and forward to the new one.

Sirius was sometimes confused with another two-headed beast called Orthrus. This was Geryon's watchdog; his job was to guard this tyrant's cattle. Heracles captured the cattle (as his Tenth Labour), killing Orthrus in the process.

In antiquity, as Homer and Hesiod were penning their stories, the Dog Star was already associated with the Sun, since the Sun enters that part of the sky in the hot summer months. While the brightest of stars, it hadn't the best of reputations in antiquity as it was said to bring sickness and death. Perhaps this was due to the fact that July and August were habitually the times of drought and disease.

The name *Sirius* may come from the Greek meaning "scorching", or it may not. Burnham's *Celestial Handbook* (as always) offers a wide background into the matter of etymology. The star is mostly thought of now as a winter star, accompanying Orion, rather than as the summer home of the sun.

Some facts about Sirius:

Although the brightest star, Sirius is rather Sun-like in size and luminescence; certainly it is no giant at an estimated 1.5 Sun diameters.

Its brightness comes from the fact that it is very close to us: at 8.56 light years away it ranks as the sixth closest star.

The star is a notable binary, but with a companion which is very dim and very close. The companion is a white dwarf, and its presence wasn't really discovered at first; it was just a hypothesis.

In 1834 Friedrich Bessel noticed a slight oscillation in Sirius's orbit. He made the calculations and predicted the existence of an unseen companion. But by his death, in 1846, the companion hadn't yet been discovered. It was only in 1862 that verification came.

This white dwarf has since been the subject of much study. Named Sirius B or *The Pup*, it is an eighth-magnitude star with an estimated radius of only 10,000 km (about twice the size of the earth). Yet its mass is nearly equal to that of our Sun's, which creates a density so high that a tablespoon full of its matter would weight over a ton.

Such a small dense object is the first phase of the collapse of the so-called *main-sequence stars*. First white dwarfs, as they continue to cool they become yellow dwarfs then red dwarfs. Finally they die completely and are known as black dwarfs.

Beta Canis Majoris is also of some interest. Its name, "Murzim" means "The Announcer", as its appearance on the horizon signifies the approach of Sirius. This is a pulsating giant that has become the prototype of a class of variable stars (see below).

The Bayer stars are quite bright, ranging from -1.5 to fifth magnitude, with a dozen stars of third magnitude or better.

Double stars:

Sirius B: The companion describes an orbit of 50.09 years. At 1 January 2000, it will have a PA of 150 degrees and a separation of 4.6".

Mu CMa is a fixed multiple binary, with components B, C, and D at these fixed spots: B: 340°, 3", C: 288°, 88.5", and D: 61°, 101".

h3945 is a gorgeous yet rather unknown binary: gold and blue. It isn't terribly difficult to find nor to resolve, and when you do find it you will keep coming back to enjoy its colours.

> The primary is a fairly bright 5.0; the companion has a visual magnitude of 6.1 and is found at PA 55° and separation 26.6".
>
> To locate the primary, first find tau CMa, which is just to the northeast of delta CMa. Now look north of tau CMa, about 1.75 degrees and very very slightly to the west of due north. You should find the

fairly bright primary with no problem. Focus carefully and study this star. Its companion should be quite visible, particularly if you enjoy clear dark skies. You will know when you find it; the colours are unmistakable.

Variable stars:

Beta CMa is a pulsating giant star, and the prototype of a small class of variables. Its variations are too slight to be noticed by the naked eye, as it changes from only 1.93 to 2.00 every 6h, 2.6s.

> This class of variable is also called the "beta Cepheid type", as this particular star was the first in this class to be discovered (in 1901).
>
> These are all young stars, with a spectra of O or B. Characteristically, they have extremely small changes in magnitude over very short periods (the longest period is ES Vul, with a period of 14h 38.4m). Interestingly, the radial velocity appears to fluctuate with the same period, often quite dramatically (e.g. more than 100 km/s). Reasons for this phenomenon are still not understood.

Other beta CMa variables are *iota Canis Majoris* and xi^1 *CMa*. Both of these stars fluctuate only about 0.04 visual magnitude; in *iota's* case, in every 1h 55m, while *xi* takes almost 5h to make the cycle.

There are no long-period Mira-type variables of any consequence in Canis Major. Indeed, unless one is studying cepheid variables, this is not a particularly fruitful area of the sky for the student of variable stars.

Deep Sky Objects:

M41 is a globular cluster easily located four degrees south of Sirius. Perhaps a hundred or so stars make up this bright group, fifty of them bright enough to be easily seen in binoculars. At the center of the group is a red giant. The group is thought to be about 2500 light years away.

NOTES

Canis Minor

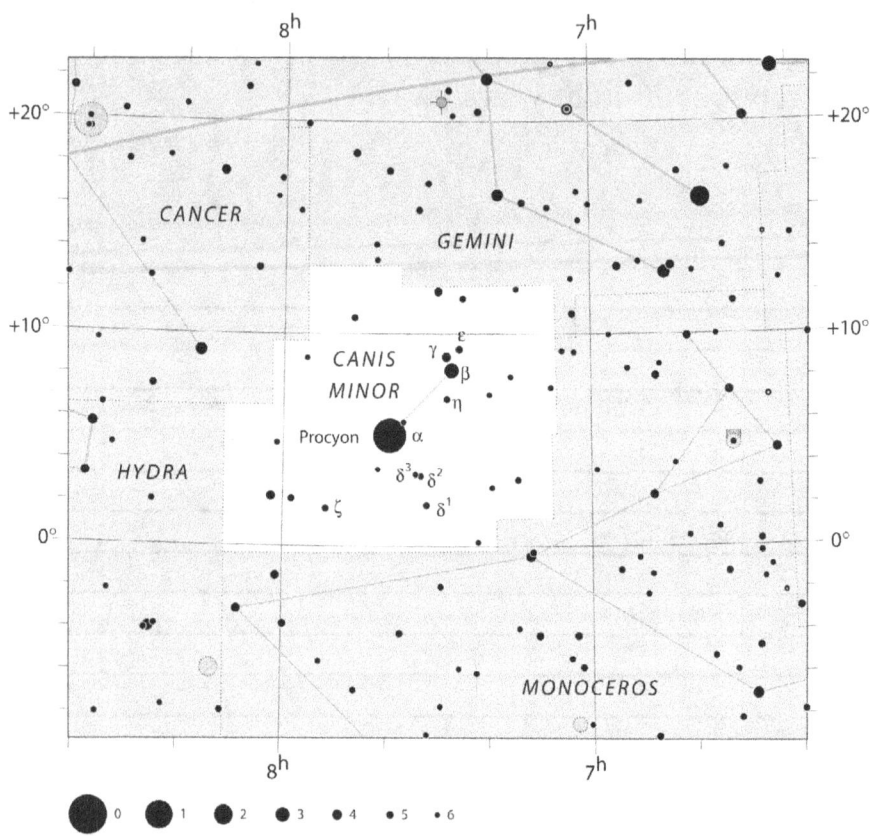

The Little Dog

Canis Minor is sometimes connected with the Teumessian Fox, a beast turned into stone with its hunter, Laelaps, by Zeus, who placed them in heaven as Canis Major (Laelaps) and Canis Minor (Teumessian Fox).
Much smaller than its mate, its only point of interest is in its principal star, *Procyon*.

> The name means "Before the Dog", referring to the fact that this star rises just before Sirius (alpha CMa). At 11.44 light years away, Procyon is nearly as close to us as Sirius (8.65 ly).

There are only nine Bayer stars, which range from 0.38 to nearly sixth magnitude.

Double stars:

> *Procyon A* and *Procyon B* form an extremely difficult binary. In fact the companion, which is a white dwarf with a diameter of only twice that of Earth's, was first seen only in 1896. Since then its orbit, which is nearly circular, has been calculated to be 40.65 years.

In 2000 the values are: 0.4, 10.3; PA 150°, separation 4.6".

Variable stars:

Beta CMi is a gamma Cas type variable, fluctuating from 2.84 to 2.92.

Deep Sky Objects:

> Canis Minor has no deep sky objects.

Capricornus

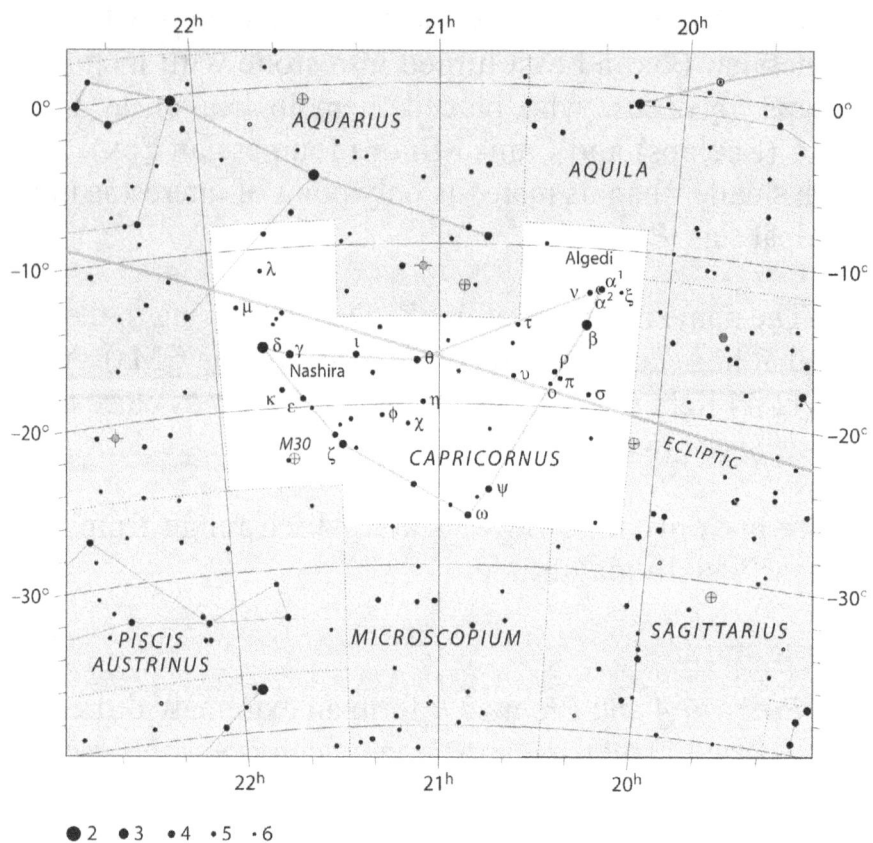

The Sea Goat

This zodiacal constellation, like Pisces, depicts the result of the sudden appearance of the earthborn giant Typhoeus. Bacchus was feasting on the banks of the Nile at the time, and jumped into the river. The part of him that was below water was transformed into a fish, while his upper body became that of a goat. From this point of view, he saw that Typhoeus was attempting to tear Zeus into pieces; he blew a shrill note on his pipes, and Typhoeus fled. Zeus then placed the new shape of Bacchus in the heavens out of thanks for the rescue. Capricornus has therefore from antiquity been represented by a figure with the head and body of a goat and the tail of a fish. It may be seen between Aquarius and Sagittarius low on the southern horizon.

Considering its importance, Capricornus is rather faint; the asterism of a horned animal isn't terribly evident, and the Bayer stars are generally third to fourth magnitude.

Alpha Capricorni is known as *Al Giedi* or *Algedi* (the goat or ibex).

> This is a double star, alpha^1 and alpha^2 Capricorni. Alpha^2 is the primary, although they only make an optical pair. Each star is however a visual binary (see below for details).

Beta Capricorni is called *Dabih*, from the Arabic *Al Sa'd al Dhabih* meaning "The Lucky One of the Slaughterers". This name indicates that the star served to signal the beginning of a winter ritual, possibly the very ones depicted on the pottery examples shown above. For the sun would have been in this constellation at the winter solstice three to four thousand years ago; were they beseeching the gods to bring back the Spring?

Delta Capricorni is the brightest star of the constellation (as well as an eclipsing binary). The Arabs called delta and nearby gamma Capricorni "The Two Friends".

Double stars:

Alpha^2 and *alpha^1 Cap* form an optical binary of yellow and orange stars: 3.6, 4.2; PA 291°, separation 378". Each star is a visual binary:

Alpha^1 Capricorni: 4.6, 9.2; 221°, 45.4".
Alpha^2 Capricorni: 3.5, 9.5; 156°, 154".

Beta Capricorni is a wide visual binary with a nice colour contast, yellow and blue: 3.1, 6; PA 267°, separation 205".

Tau Capricorni is a visual binary with a 95 year orbit: 5.5, 7; PA 107°, separation 0.4".

Variable stars:

The only variable worth noting is delta Capricorni, which is an eclipsing binary (2.81-3.05) with a period of 24h32m47.2s.

Deep Sky Objects:

> The only Messier in the constellation is *M30 (NGC 7099)*, a globular cluster about 40,000 light years away. It has a very concentrated centre, with a number of star chains (or strings) from the centre to the periphery.
>
> *M30* is 3° ESE of zeta Capricorni.

NOTES

Carina

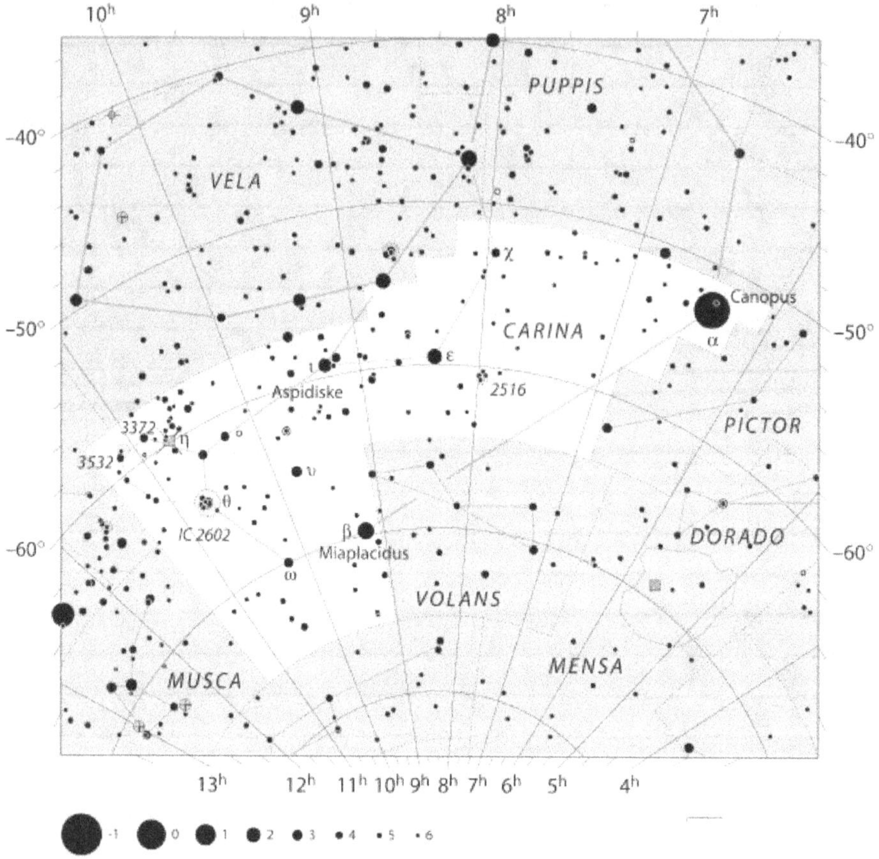

The Keel

When the twins Castor and Pollux went off with Jason and the rest of the Argonauts, they sailed in the Argo, a ship built by Argus. This ship, equipt with fifty oars and manned by fifty of the best men of Greece, sailed to Colchis, which was at the eastern shore of the Black Sea.

After many adventures Jason (with Medea's help) stole the fleece from the dragon and they all sailed back home.

Athene is said to have commemorated the event by placing their ship, Argo Navis, in the sky as a giant constellation below and east of Canis Major. What is known is that Edmund Halley's catalog of the southern stars, Catalogus stellarum australium (1679), introduced Argo Navis to the world.

In 1763 Nicolas Louis de Lacaille's posthumous work *Caelum australe stelliferum* gave us most of the constellations we now know in the southern hemisphere. Lacaille divided the gigantic Argo Navis into three constellations: Carina (the Keel), Puppis (the Stern, or Poop deck), and Vela (the Sail). To this day the Bayer letters (Greek letters) are divided among these three. By far the most interesting of these three is Carina.

Carina is home to Canopus (alpha Carinae), the second brightest star in the heavens.

> Its name is said to come from the pilot of the fleet of ships of King Menelaus. It was this king of Sparta who rallied the men of Sparta to fight for Helen of Troy; his prize was Helen herself, who became his queen. Of Canopus, the pilot, it is said that he died in Egypt after the fall of Troy.
>
> Canopus -- the star -- was known in antiquity as the

Star of Osiris and worshipped in many ancient cultures. This was the star that Posidonius used in Alexandria, circa 260 BC, as he became the first person to plot out a degree of the Earth's surface.

While the brightest star in the southern hemisphere, Canopus is not visible to anyone living above latitude 30 degrees north. Thus Europe north of Lisbon cannot see the star, and for North America the star is visible only for those living south of a line drawn between San Francisco and Washington D.C., depending on local topography of course.

For inhabitants of the southern hemisphere, Canopus announces the beginning of summer, for it culminates on December 27.

Canopus is a supergiant around thirty-five times the diameter of the Sun. Prior to the Hipparcos satellite, the distance of Canopus was difficult to calculate, with estimates ranging from 160 to over 1200 light years. Hipparcos has calculated the distance at 313 light years (96 parsecs). The star has a luminosity of over 12,000.

As pointed out above, there are only a few Bayer stars, as this constellation is only part of an originally much larger one.

The most interesting object of the constellation is *Eta Carinae*: a mystery star which changes its magnitude very irregularly, from a brilliant -0.8 in 1843 to a rather dim -7 in the mid 1870s. Its present visual magnitude isn't much brighter, at only 6.21.

The star's absolute magnitude has been difficult to assess. If we take the distance to be 2000 parsecs, as some authorities would have it, then we arrive at an absolute magnitude of -5.3 (and a distance of 6500 light years). On the other hand Burnham gave an absolute magnitude of -3.3, and a distance of 2600 light years.

The star is considered to be either a very young one, not yet on the main sequence, or a very old one, approaching its eventual demise. At the present time the latter view seems prevalent. When it finally does die, it might create one of the brightest supernovae ever seen.

Eta Carinae is associated with the Keyhole Nebula (see below)

Double stars:

Upsilon Carinae is easily resolved, a pleasant binary of two white stars (visual magnitudes 3.0, 6.0), 127° and separation of 5.0".

Variable stars:

Chi Carinae is a cepheid variable, from 3.46 to 3.475 every two hours and twenty-five minutes.

R Carinae is a Mira-type red giant ranging from 3.9 to 10.5 every 308.71 days. In the year 2000 the maximum should be reached in the last week of January.

ZZ Carinae ("el") is an unusual cepheid with variations that are quite noticeable to the naked eye. From its maximum of 3.3 it slowly dims over a three-week period to about 4.0, then it takes only seven days to achieve its brightest again, before the slide begins all over again.

Deep Sky Objects:

NGC 2516. A very nice open cluster of perhaps a hundred stars, located fifteen degrees SE of Canopus. with a red giant at the centre. It's estimated at 1200 light years away.

NGC 3372, The Keyhole Nebula. A diffuse nebula of great complexity and beauty. While the nebula is composed of brightly glowing gas, there are darker areas which serve to break the nebula into individual islands. The most dramatic of these darker areas has been labelled the *Keyhole* because of its shape. Eta Carina is found in this nebula.

NGC 3532. A spectacular cluster of four hundred or so mostly bright, sparkling white, class A stars. John Herschel thought this was the finest cluster he'd ever seen. It's located three degrees WNW of eta Carinae.

IC 2602 is a group of thirty or so stars some 700 light years away; *theta Carinae* is the brightest member.

NOTES

Cassiopeia

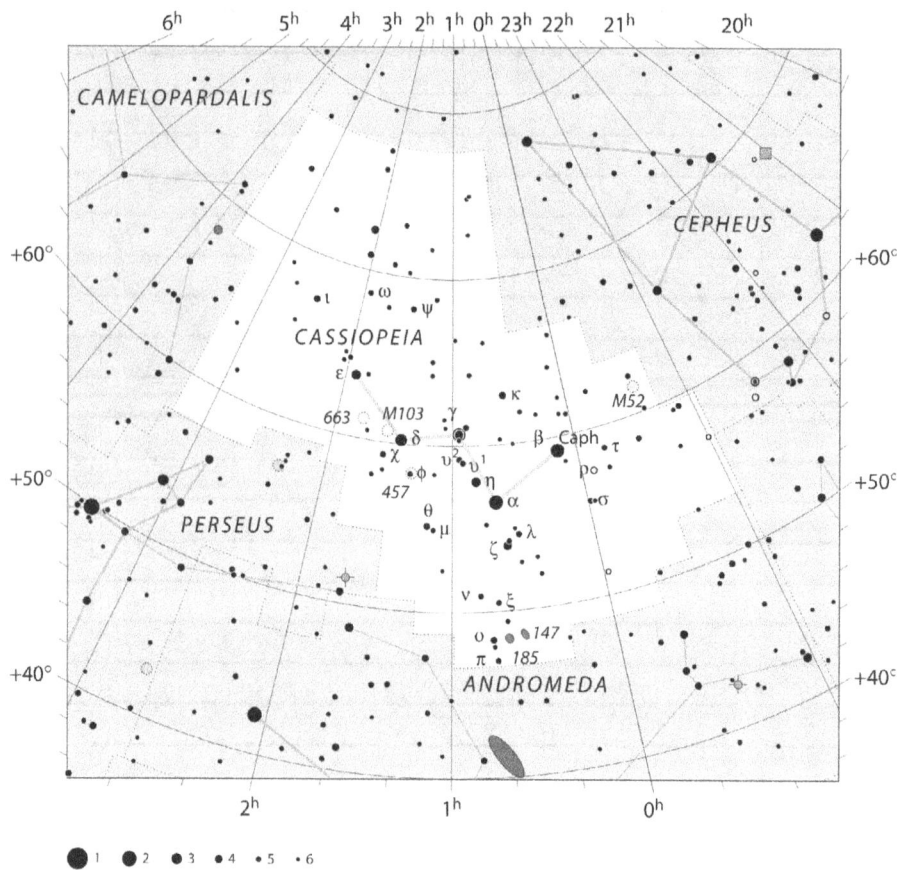

The Queen of Ethiopia

Cassiopeia was the beautiful wife of Cepheus, king of Ethiopia, and the mother of Andromeda. She is most famous in connection with the myth of her daughter, Andromeda. The queen made the mistake of bragging she was more lovely than the Nereids, or even than Hera herself. The goddesses were, needless to say, rather insulted, and went to Poseidon, god of the sea, to complain. Poseidon promptly sent a sea monster (possibly Cetus?) to ravage the coast. The king and queen were ordered to sacrifice their daughter to appease Poseidon's wrath, and would have done so had Perseus not arrived to kill the monster in the nick of time. As a reward, the hero was wedded to the lovely Andromeda.

By most accounts, Cassiopeia was quite happy with the match. In some versions of the myth, however, the queen objects to the marriage and is turned to stone when Perseus shows her the head of the Gorgon Medusa.

Although she was placed in the heavens by Poseidon the sea-god saw fit to humiliate her one final time (and for all eternity). He placed her so that she is seated on her throne, with her head pointing towards the North Star Polaris. In this position, she spends half of every night upside-down.

The asterism clearly shows the chair upon which Cepheus's queen sits. The Bayer stars are generally third and fourth magnitude, with the exception of the first four stars which make up the "chair".

Cassiopeia has many fine binaries, a few variables of note, and several interesting deep sky objects.

Double stars:

Gamma Cassiopeiae has a faint companion, made doubly-difficult to see because of the brightness of the primary: 2.5, 11; PA 252° and separation 2.3".

Eta Cas is a fine binary with colour contrast, yellow and red. Some observers see them as more gold and purple.

> The companion orbits every 480 years. Present values are: 3.4, 7.5, PA 315°, separation 12.7".

Lambda Cas has two nearly equal stars: 5.5, 6; PA 179°, 0.5".

Iota Cas is a triple system, with AB a visual binary with an orbit of 840 years.

> **AB: 4.6, 6.9; presently at PA 231° and separation 2.5".**
> **C: 8, PA 114°, 7.3".**

Omicron Cas has a faint companion: 4, 11; 302°, 33.6".

Phi Cas is another multiple system, with rather wide components. The binary lies on the edge of *NGC 457* (see below).

> **AB: 5, 12; 208°, separation of 48.6".**
> **C: 7; 231°, 134".**

Sigma Cas: 5.0, 7.1; 326°, 3".

> *Struve 3062* is a visual binary with orbit of 106.8 years: 6.4, 7.5; presently 322° and separation 1.5".

Variable stars:

Beta Cas is a delta Sct: 2.25-2.31 with period of 0.104 days (2h 30m 11.5s).

Gamma Cas is a prototype of an important class of variable.

> "Gamma Cas" variables are B stars, very rapidly rotating subgiants or even dwarfs with emission spectra. The variation in magnitude is typically quite small.
>
> The biggest exception is gamma Cas itself, which has a range of 1.5 to 3.0 with a sporadic period, roughly every 0.7 days.
>
> Other stars in this class include zeta Tau and BU Tau ("Pleione"), mu Cen, lambda Pav, and epsilon Cap.

Iota Cas is an alpha CVn type variable: 4.45 to 4.53 every 1.74 days.

Omicron Cas is a gamma Cas type variable, ranging from 4.5 to 4.62.

R Cas is a Mira type variable with a period of 430.46 days, ranging from 4.7 to 13.5. A maximum should occur in the last week of August in the year 2000.

Deep Sky Objects:

Cassiopeia has two Messier objects and several other star clusters of interest.

> *M52 (NGC 7654)* is an open cluster of about 120 stars. It's found 6° NW of rho Cas. Burnham gives the best method of finding the cluster: draw a line from alpha Cas to beta Cas, then continue this line, doubling its

length. The cluster is just past the end point, about another quarter-length.

M103 (NGC 581) is another open cluster, with about forty stars. It's 1° NE of delta Cas, or 1.5° due north of chi Cas.

NGC 457 is an open cluster about 4° SE of gamma Cas. The star phi Cas is considered a part of this cluster. This star is one of the most luminous known, with at least 200,000 times the light of the sun.

NGC 7789 is a rich open cluster of perhaps a thousand stars. It's 3° SW of beta Cas, lying just between rho Cas and sigma Cas.

NOTES

Centaurus

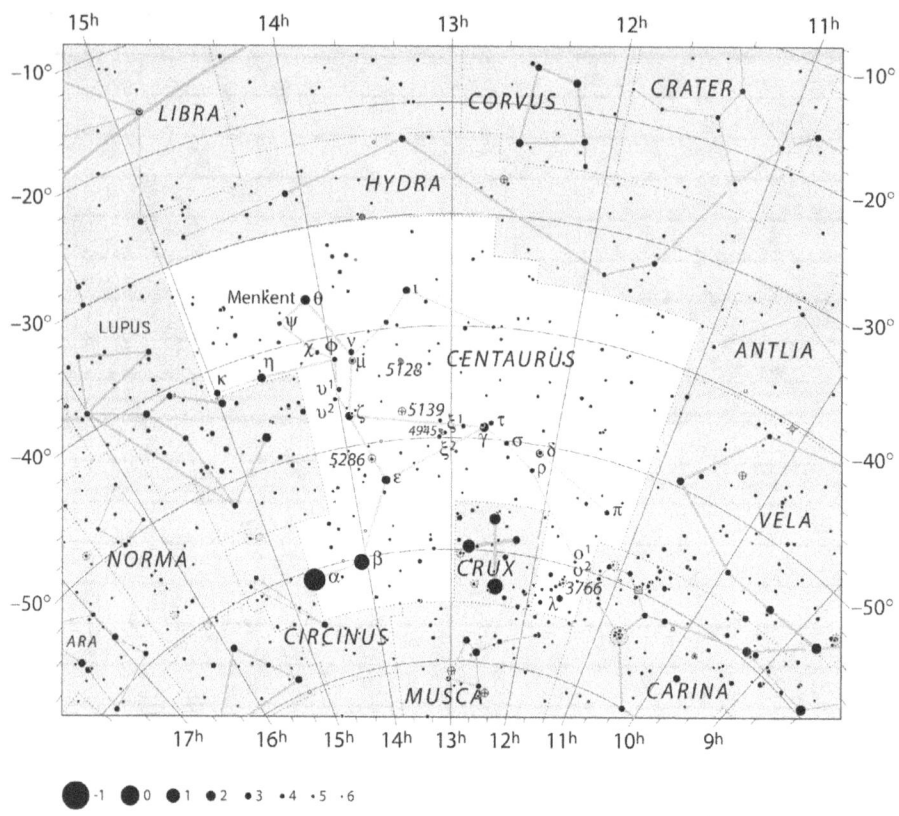

The Centaur

₡entaurus is one of several constellations that deal with the Labours of Herakles.

In the Fourth Labor, Heracles' assignment was to bring back a rampaging wild boar that was bringing death and destruction to the inhabitants of the northern part of the Peloponnesian peninsula. On his way, he stops to visit a friend of his, a Centaur named Pholus.

Centaurs were half-men, half-horse, who had all descended from Ixion and Nephele (who was in fact a cloud, shaped by Zeus to resemble his wife Hera).

Centaurs were featured in a number of Greek myths, but by and large remained on the periphery of Greek fable.

As Heracles finishes the sumptuous meal provided by Pholus, he then has the effrontery of opening the Centaurs' private wine cask, meant for them alone. The rest of the Centaurs catch the odor of their wine, wafting across hill and dale, and they become enraged.

Gathering up huge boulders, ripping out trees to use as clubs, and arming themselves with axes, the Centaurs advance on the dinner party.

Pholus takes fright, so the battle is left to Heracles. After repulsing a number of Centaurs single-highhandedly, Heracles then chases the rest of them to the cave of their king, Cheiron.

Herakles shoots an arrow at one fleeing Centaur (Elatus by name), but it passes through his arm and strikes Cheiron on the knee. You may recall that Heracles' arrows were all dipped in poison, so each was fatal, no matter how slight the wound. Cheiron was a great friend of Heracles, and our hero is devastated. He tries to assist Cheiron, but there is nothing to be done.

Cheiron was immortal, so the poison couldn't kill him, only cause great pain that would last through eternity. He descends to the depths of his cave, his screams of agony echoing throughout the cavernous walls.

Eventually Prometheus takes pity on the long-suffering king of the Centaurs, and offers to take over Cheiron's immortality, if Zeus would agree. Zeus does agree, so Cheiron's agony finally comes to an end, and Zeus places the great king of the Centaurs in the heavens.

Back to the previous battle. The Centaur Pholus looks over the dead and dying and wonders how Heracles' arrows could be so fatal. He plucks one arrow out of a body and looks at it, but it slips through his fingers and strikes him on the foot, killing him instantly.

Heracles hears of the tragedy and returns to bury his friend, at the foot of the mountain that bears his name: Mt Pholoe.

This high plateau region in the interior of the peninsula is just up the road from Olympia. Now called Pholois, this is where the Centaur stories of antiquity originated.

It is said that Zeus had held Pholus in very high regard, and therefore also put his likeness in the heavens. Thus the constellation Centaurus represents two Centaurs: Pholus and Cheiron.

The fact that two Centaurs are linked with the constellation is no accident. The earliest extant artifact showing the likeness of a Centaur is a piece of Mycenaean jewellery which shows two centaurs together: half-men, half-horse, facing each other and dancing, similar to satyrs.

Centaurus is one of the largest constellations with a clearly discernible asterism: the huge form faces east, with a sword waving menacingly toward Lupus the Wolf on the west.

The constellation has an almost complete list of Bayer stars except for *omega*, which isn't a star, but a well known globular cluster, NGC 5139 (see below).

The front hooves (or feet, if you wish) are formed by two bright stars: alpha and beta Centauri, known also by the Arabian names of Wazn and Hadar.

Alpha Centauri is best known by the name "Rigil Kentaurus", or the Centaur's foot. This is a triple system, three stars which are the closest to our own Sun.

> *Alpha^1* and *alpha^2 Centauri* form a noted binary (see below). They are 4.393 light years away, and each is approximately the size of the Sun.
>
> The closest star is actually *alpha^C*, known as *Proxima Centauri*. This is a red dwarf of visual magnitude 11.01 and distance 4.221 light years.
>
> > The star has a diameter of about 65,000 km (40,000 miles), or about five times that of the Earth. It is a great distance from the other two (perhaps as far as a sixth of a light year away) and the orbit's period is estimated to be hundreds of thousands of years; Burnham suggests "perhaps ... half a million years".

Proxima Centauri is a flare star and is therefore also known by its variable designation of *V645 Centauri*. See below for details.

Beta Centauri (Hadar) is the tenth brightest star in the heavens, at 0.61 visual magnitude (which is actually the combined values of its two components). It's 525 light years distant and is a rather difficult visual binary (see below).

Double stars:

Alpha^1 and *alpha^2 Centauri* form a wide double with an orbit of 79.92 years: -0.04, 1.2. The 2000.0 values are PA 222° and separation 14.1".

Beta Centauri is a difficult double because of the primary's brightness compared to the companion: 0.58, 3.95; PA 251°, separation 1.3". The orbit has not been calculated, but is thought to be at least several hundred years.

Gamma Centauri is a visual double of two nearly identical stars, with orbit of 84.5 years: 2.9, 2.9. The 2000.0 values are a PA of 347° and separation 1.0".

Eta Centauri is a binary with very faint companion: 2.3, 13; PA 270°, separation 5.6".

Kappa Centauri also has a faint companion: 3.1, 11; PA 82°, separation 3.9"

Variable stars:

Alpha^C Centauri (V645 Centauri) is a flare star. That is, its visual magnitude may change rapidly, perhaps taking only several seconds to change its magnitude.

The prototype of this kind of variable is UV Ceti, which has been known to change 3.5 magnitudes within seven seconds!

> Since these stars are extremely dim, only the closest ones have been investigated. There are twenty or so such stars within twenty light years of our Sun; they all have an M4.5-M6.5 spectra. Only two have visual magnitudes brighter than 10.

R Centauri is a Mira type variable, 5.3 to 11.8, with 546.2 year period. In 1999 the maximum should arrive in mid-February; the same in 2002.

> As the chances of seeing this star at its maximum are therefore not very likely, you might find Burnham's finder's chart (p. 559) of some use.

Deep Sky Objects:

NGC 5139, also known as *omega Centauri*. This globular cluster is usually described as the finest in the heavens. It's so bright and compact, Bayer thought it was a hazy star, and named it omega.

The cluster is estimated to be from 15000 to 25000 light years away, and may be comprised of over a million stars.

> The cluster lies between gamma and zeta Centauri, about five degrees west of zeta.

Cepheus

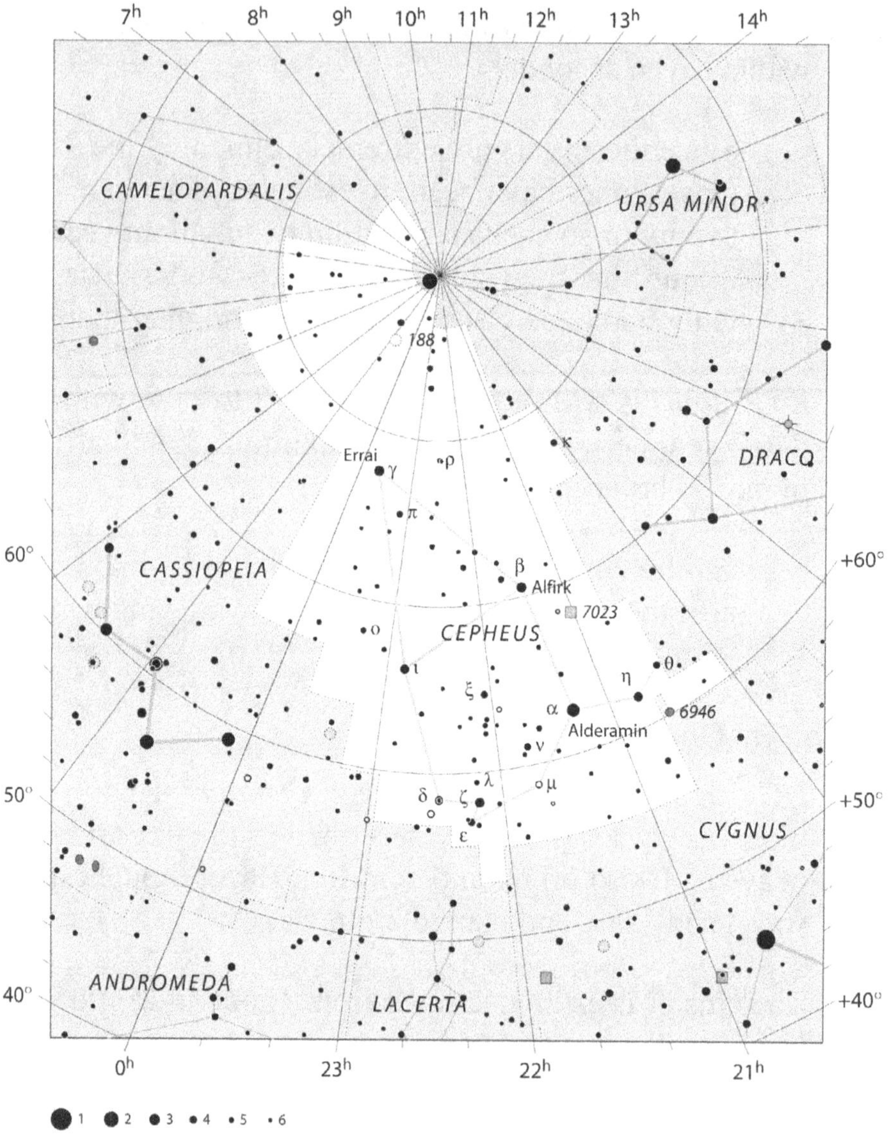

The King of Ethiopia

𝕮epheus is the name of two mythological kings. One was the son of Aleus, from Arcadia. He would become the king of Tegea (a community on the Peloponesian peninsula), would father twenty children, and would sail with Jason as an Argonaut.

The other Cepheus was the son of Belus, king of Egypt (who was himself the son of Poseidon). This Cepheus grew to become the King of Ethiopia (or Joppa). He married Cassiopeia and they had a daughter Andromeda. (Yes, the *whole family* eventually winds up in the heavens.)

Cassiopeia was incredibly beautiful but immensely vain. She was also proud of her daughter's beauty. In fact she continually boasted that the two of them were more beautiful than any of the fifty sea nymphs who attended Poseidon's court.

These nymphs (the Nereids) complained to Poseidon, who felt he had to defend his own reputation. So he sent a flood to devastate Cepheus' kingdom. The oracles told Cepheus that in order to save his people he must sacrifice his daughter to a great sea monster: Andromeda was tied to a rock along the coastline, dressed only in her jewellery. The monster would be along in due time to take his prize.

At that moment Perseus came flying by. He had just killed the Gorgon Medusa and was carrying the severed head back to Athene.

> Medusa was one of the three Gorgon sisters; they were once very beautiful but Medusa slept one night with Poseidon, in Athene's temple. This infuriated Athene so much she turned Medusa's hair to snakes and turned her into a terrifying monster with huge teeth and claws.

One look from Medusa would turn the viewer to stone.

With the assistance of Athene, Medusa's sworn enemy, Perseus tricked Medusa by looking at her reflection. He then sliced off her head (Pegasus and a warrior named Chrysaor sprang fully-formed out of Medusa's dead body).

So Perseus arrives at the scene and has a quick chat with Cepheus and Cassiopeia; it is agreed that should he rescue their daughter, he can marry her. So he skims across the water and his shadow on the waters confuses the monster, which Perseus then beheads.

Far from delighted, Cepheus and Cassiopeia balk at their daughter marrying Perseus. However Andromeda insists, so the marriage ceremony is performed on the spot.

Halfway through the ceremony Agenor, a family relative, shows up and claims Andromeda as his bride. (It's pretty certain now that Cassiopeia put him up to it.) Understandably, this angers Perseus and a great battle breaks out. However, outnumbered, Perseus has to resort to desperate measures. So he shows Medusa's head, instantly turning everyone to stone, including Andromeda's parents.

Poseidon then put Cepheus and Cassiopeia into the heavens, but with a twist: he made the vain Cassiopeia spin around on her chair, spending half the year upside down. As for Cepheus, Poseidon gave him a number of medium sized stars that go to make his square face with a pointed crown.

If not very bright, the constellation is still quite noticeable, just to the west of Cassiopeia's chair. The stars are mostly third and fourth magnitude.

The constellation has numerous binaries, several significant variables, and a few interesting deep sky objects.

Alpha Cephei is known as *Alderamin* ("The Right Shoulder"). In another 5500 years this will become the Pole Star.

Beta Cephei is called *Alfirk* ("The Herd"). This is a visual binary as well as a variable (see below).

Gamma Cephei is *Er Rai* (Shepherd). Long before Aldemarin becomes the Pole Star, this one will assume that title (around 4000 AD).

Delta Cephei is a prototype for one of the more significant types of variables (see below). The star is also a very fine binary with a colour contrast.

Mu Cephei is a brilliantly coloured star, a deep red which moved William Herschel to call it "The Garnet Star". The colour depends on the size of one's telescope; the larger scopes bring out an orange element. It is also a semiregular variable (see below).

Double stars:

The constellation has a number of very fine binaries, some quite easily resolved by small scopes, others rather more difficult.

Beta Cephei is a blue giant with a faint companion

easily resolved: 3.2, 8; PA 250°, separation 13.6"

Delta Cephei is a fixed double, a yellow giant with blue companion: 3.8, 7.5; PA 191°, separation 41".

Xi Cephei is considered to be the most attractive binary in Cepheus, a blue-white primary and yellow (or reddish) companion that orbits every 3800 years: 4.4, 6.5; PA 275°, separation 8.2". 6.5;

Kruger 60 is a famous binary only 12.9 light years away, comprised of two red dwarfs. Observers had reported seeing flare-ups on the surface of the companion, which is orbiting the primary every 44.6 years.

AB: 9.8, 11.4; presently the companion is at PA 109° and separation 3.2".

The binary is less than one degree SSW of delta Cephei. Burnham has a finder's chart (p. 600).

Struve 2816 is a multiple binary, a very attractive triple:

AC: 6.3, 8.1; 121°, separation 11.7"
D: 8.0; PA 339°, 19.9"

In the same field are *Struve 2813* and *Struve 2819*, all centred in the middle of the large but faint diffuse nebula IC 1396, just south of mu Cephei.

Variable stars:

Beta Cephei is the prototype of a class of pulsating

variables. These are hot, luminous, and very massive stars with a specral type of O9 to B3. The group is also called 'beta Canis Majoris' stars, for this star too is a member of the group. Indeed, it is the brightest member of the group.

The variations in visual magnitude are of very small amounts, barely as much as a quarter of a magnitude. Beta Cephei varies only 0.04 magnitude from its typical brightness of 3.2 and a period of 4 hours 34 minutes.

Delta Cephei is also the prototype of a class of variables.

Delta Cephei variables are immensely useful stars. Studies of the period of pulsation and the apparent magnitude led investigators to devise a method of gauging the distance to outlying galaxies (the so-called *period-luminosity relation*).

> *Delta Cephei* varies from 3.48 to 4.37 every 5 days, 8 hours, 47 minutes, 31.9 seconds. This means its maximum brightness can be calculated with a reasonable degree of accuracy.
>
> After the maximum is reached, a gradual diminuition of magnitude occurs over the next three days, only for the star to again increase over the following two days until it again reaches its maximum.
>
> Comparing its magnitude with zeta, just to the west, will tell you if it has reached its brightest. Zeta has a visual magnitude of 3.4, while delta Cephei varies from 4.4 to 3.5. That is, for most of its cycle, it will be rather more dim than zeta, but as it reaches its maximum, it should appear to have a magnitude quite similar to its neighbour.

Mu Cephei is a semiregular supergiant that also varies roughly from 4.5 to 3.5, with a very long period: 730 days.

S Cephei is a long-period Mira-type variable, 7.4-12.9, with a period of 486 days. In the year 2000 the maximum should occur in late January or early February.

> *T Cephei* is also a long-period Mira-type variable, ranging from 5.2 to 11.3 every 388 days. It should reach a maximum in the last week of December of 1999.

Deep Sky Objects:

NGC 188 is a faint globular cluster of 150 stars. Its significance lies in the fact that it is extremely old: it is estimated to have formed ten to twenty billion years ago.

> The easiest way to find it is drop down four degrees from the Pole Star, toward the star called "2 UMi", which is nevertheless in Cepheus. The cluster is just to the southwest of this star.

NGC 6939 is a fine open cluster of about eighty stars in a very rich field that includes NGC 6946, a face-on spiral galaxy.

> The cluster is found about 2.5 degrees south of theta Cephei, or about two degrees SW of eta Cephei (thus it forms a rough triangle with these two stars).

NOTES

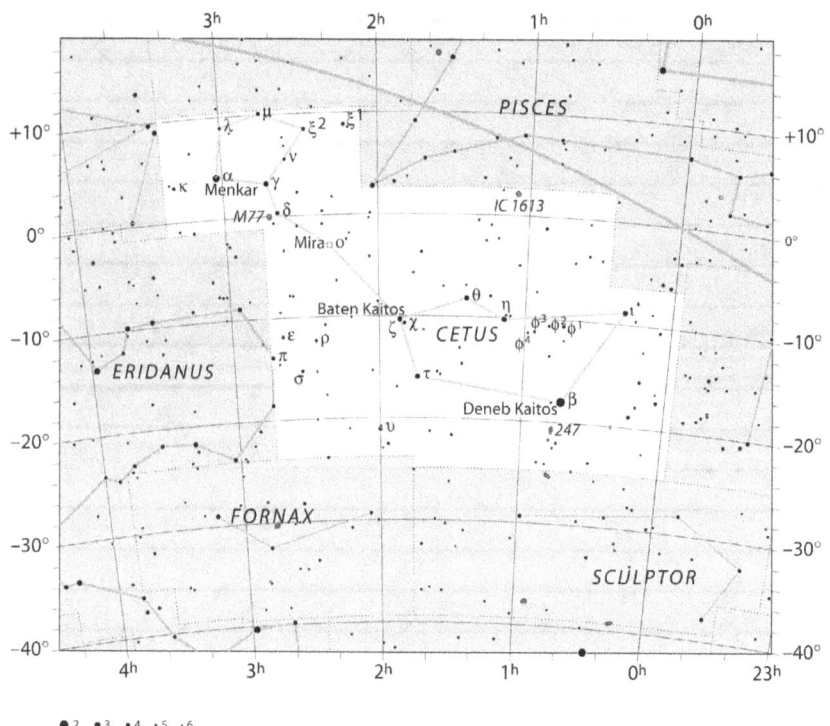

The Whale

Cetus deserves mention because some say the constellation represents the sea monster sent to Ethiopia as punishment for the boasting of Queen Cassiopeia. The monster nearly kills Andromeda, daughter of Cassiopeia and Cepheus, but is itself killed by the hero Perseus.

More frequently, though, Cetus is represented as a whale, which implies no connection to the Andromeda myth--though it certainly is possible that the ancients perceived whales as monstrous creatures. Either way, the constellation is appropriately a large one, and is relegated to the southern sky--far from Andromeda, Cepheus, and Perseus.

Cetus' stars are quite faint, but there are a few well known stars here. Such as UV Ceti, which is actually a pair of red dwarfs 9 light years away, and Mira (omicron Ceti) - perhaps the most famous variable.

Double stars:

Gamma Ceti (Struve 299) is perhaps the finest binary in Cetus. Some observers find a colour contrast, yellow and blue: 3.5, 7.3; PA 294°, separation 2.8".

Nu Ceti: 5.0, 9.5; PA 83°, separation 8".

Struve 186 is a close binary of two equal stars, with an orbit of 170 years: 7.0, 7.0; currently PA 60° and separation 1.1".

beta 395 is a very rapid visual binary, with orbit of 25 years: 6.3, 6.4; presently PA 289° and separation 0.5".

Variable stars:

Omicron Ceti, better known as *Mira, "The Wonderful"*. This is the prototype of long period variables and one of the earliest variables ever discovered.

A Dutch astronomer, David Fabricus, considered it a nova in 1596. It wasn't remarked on again until 1603, when Bayer put it in his catalogue under the name *omicron*. He apparently had stumbled across the star at one of its maximums. Later attempts to find the star failed, until once again it made an appearance. Since mid-seventeenth century the star has been studied closely.

> The star has a potential range from as dim as 10 to as bright as 2.0, although it usually reaches a maximum visual magnitude of from 3 to 4. The average period is 331.96 days and the star only maintains its maximum for a few weeks, before rapidly losing its brilliance.
>
> In the year 2000 the maximum should occur in September. But the period may change slightly. It has been known to vary from as long as 353 days to as short as 304 days. Burnham (p. 636) has a finder's chart.

UV Ceti is the prototype of a classification of variables known as flare stars. UV Ceti is actually component B of a binary system composed of two red dwarfs, both having a visual magnitude of only 15.5. Combined, their magnitude is about 12.

> Every ten hours or so UV Ceti suddenly jumps in magnitude. In just a few seconds it will increase

by three or four magnitudes, even five magnitudes on occasion. Then over the next five to ten minutes the star settles back down to its former dim self.

Being so faint, flare stars must be very close to the solar system to be noticed; indeed many of them are within fifteen light years from us. UV Ceti is 8.4 light years away. (The closest flare star is V645 Centauri, better known as *Proxima*, or alpha^C Centauri, which is 4.22 light years away.)

The two stars which make up this binary are among the least massive stars known, with each component having about a tenth of the sun's mass.

The binary's period is about 26 years and the separation remains roughly 2".

To find UV Ceti first locate tau Ceti, then the binary h 2067 (see above). UV Ceti is in the same viewing area, just half a degree to the southwest of h 2067. Burnham (p. 642) has a finder's chart.

Deep Sky Objects:

M77 (NGC 1068) is a small spiral galaxy seen face on, one of the so-called Seyfert gallaxies, which means it has a radio source - a feeble example of a quasar.

M77 is about 50 million light years away, and is found one degree SE of delta Ceti.

NGC 247 is a large and fairly bright spiral galaxy with compact nucleus.

Chamaeleon

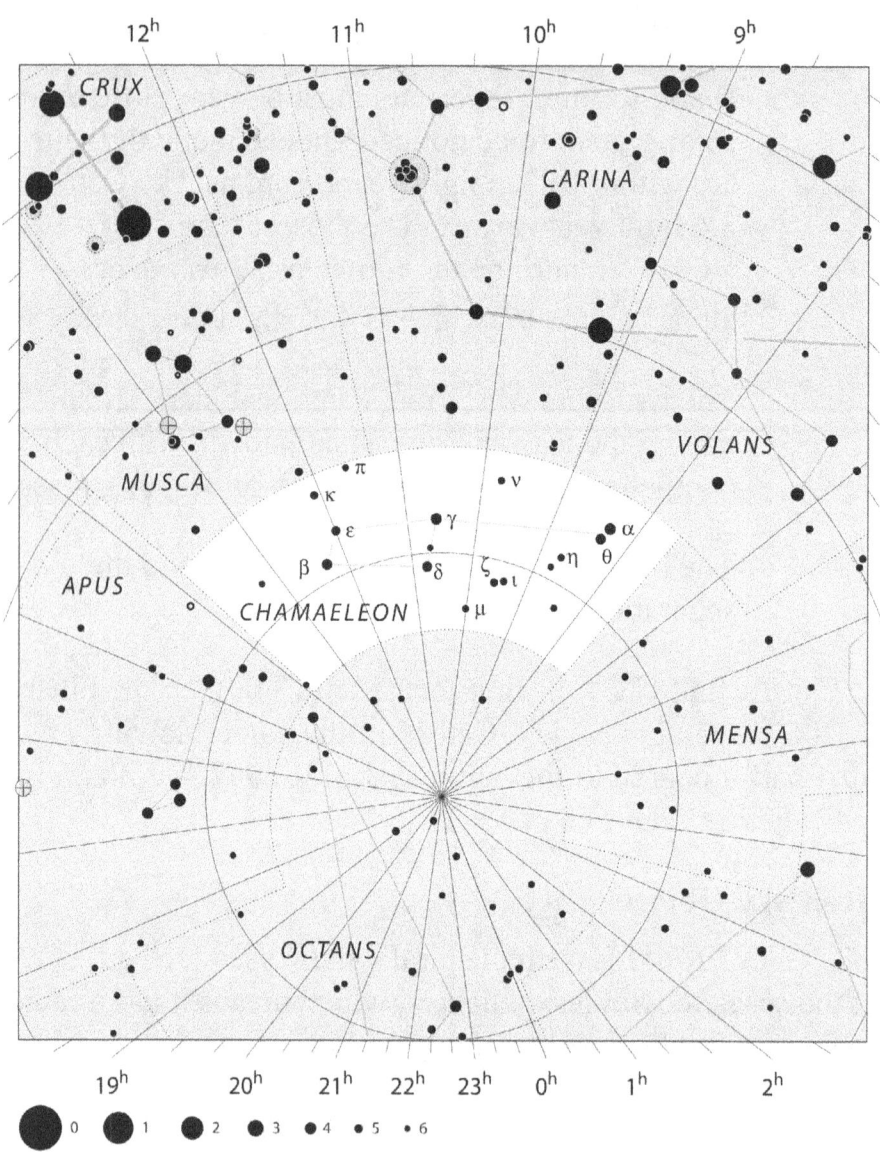

The Chamaeleon

Chamaeleon is one of a dozen constellations introduced by Johann Bayer in 1603 for his star atlas Uranometria. Like most of these, Chamaeleon is far to the south. In fact, its stars are circumpolar to residents of the Southern Hemisphere.

The asterism is supposed to represent a chamaeleon. Apparently the animal has changed itself into a rhomboid.

There are only a handful of Bayer stars, and these are generally forth and fifth magnitude.

There are a couple of binaries, a Mira-type variable, and one deep sky object of some interest.

Double stars:

Delta^1 Chamaeleontis is a close binary of nearly equal stars: 5.4, 6.5; PA 76°, separation 0.6".

Epsilon Chamaeleontis is also a close binary: 5.5, 6.0; PA 188°, separation 0.9".

Theta Chamaeleontis is a very wide binary: 4.4, 7.7; PA ?, separation 31".

Variable stars:

R Chamaeleontis is a Mira-type variable with period of 334 days, and a range of 7.5-14.

Deep Sky Objects:

NGC 3195 is a fairly bright planetary nebula located just midway between delta and zeta Cha.

Circinus

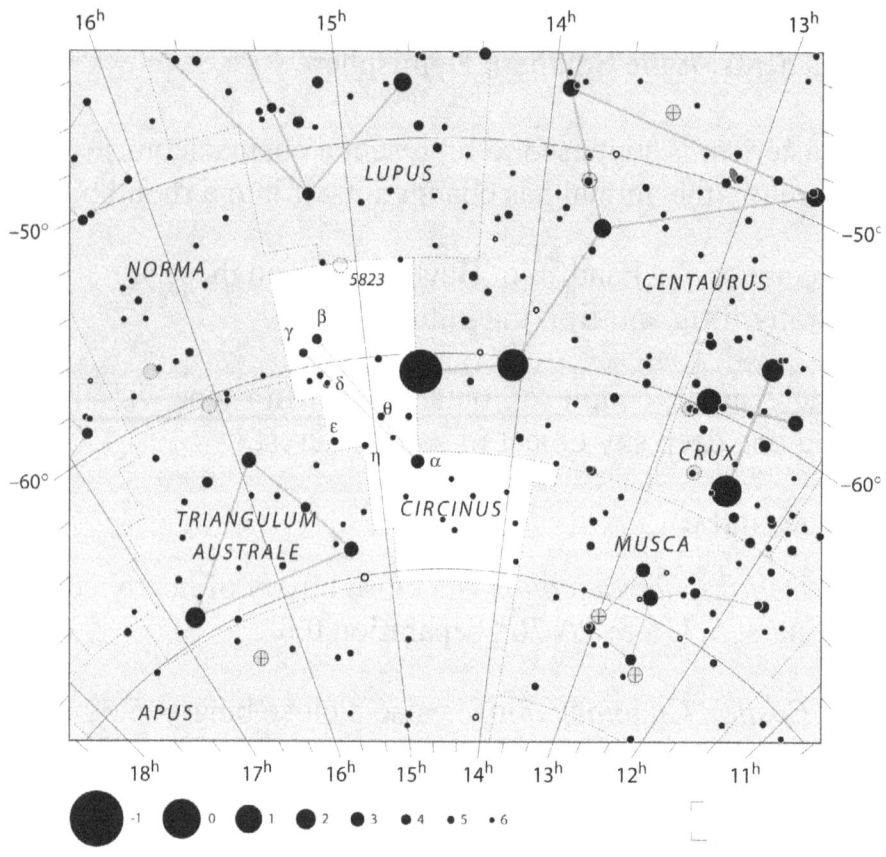

The Compass

Circinus, The Compass, is a Southern Hemisphere constellation introduced by Nicolas Louis de Lacaille in the mid-eighteenth century. Only seven Bayer stars are brighter than sixth magnitude.

Double stars:

Alpha Circini is a visual double with faint companion: 3.2, 9; PA 232°, 15.7".

Gamma Circini is a very close double of contrasting stars (blue and yellow): 5, 5; PA 49°, separation 0.9"

Variable stars:

Circinus has two variables among the Bayer stars:

Alpha Circini is an alpha CV type variable: 3.18-3.21.

Theta Circini is a gamma Cas variable: 5.02-5.44.

Deep Sky Objects:

Circinus has no deep sky objects of any interest.

NOTES

_____
_____
_____
_____
_____
_____
_____

Columba

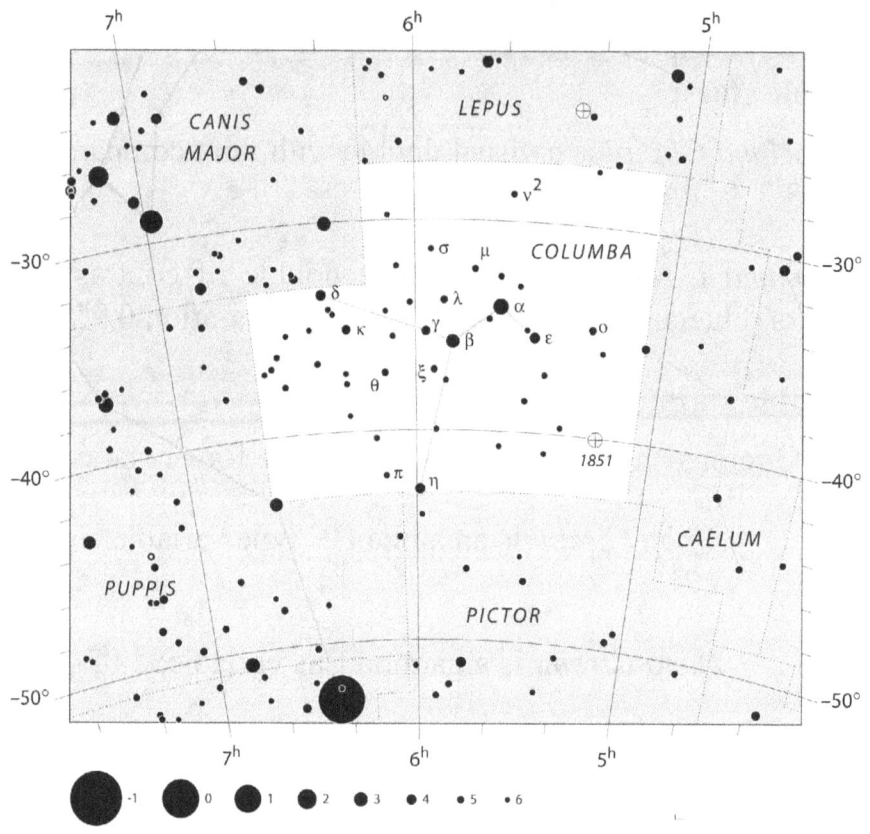

The Dove

Columba, "The Dove", may refer to the bird the Argonauts sent ahead, to help them pass the narrow strait at the mouth of the Black Sea.

However, early atlas makers called it "Columba Noae", referring to the story of Noah and the Ark, and they depicted the dove carrying an olive branch in its beak.

The constellation's Bayer stars are not complete, and mostly are in the fourth and fifth magnitude range.

Double stars:

Columba has no binaries of interest. *Phaet (alpha Col)* is only an optical (with an 11- magnitude companion), although Tirion's *Sky Atlas* marks it as a true binary.

Variable stars:

The constellation has a number of Mira-type variables; of these T Col is the brightest.

R Col varies from 8.0 to about 14 every 327 days. It's three degrees NE of mu Col.

T Col varies from 6.7 to 12.6 every 225 days. It's located one degree NNE of omicron Col.

Deep Sky Objects:

Of the few deep sky objects in Columba, NGC 1851 is perhaps the most interesting. This is a globular cluster of rather faint stars in a remote section of the sky, to the southwest, eight degrees SW of alpha Col.

Coma Berenices

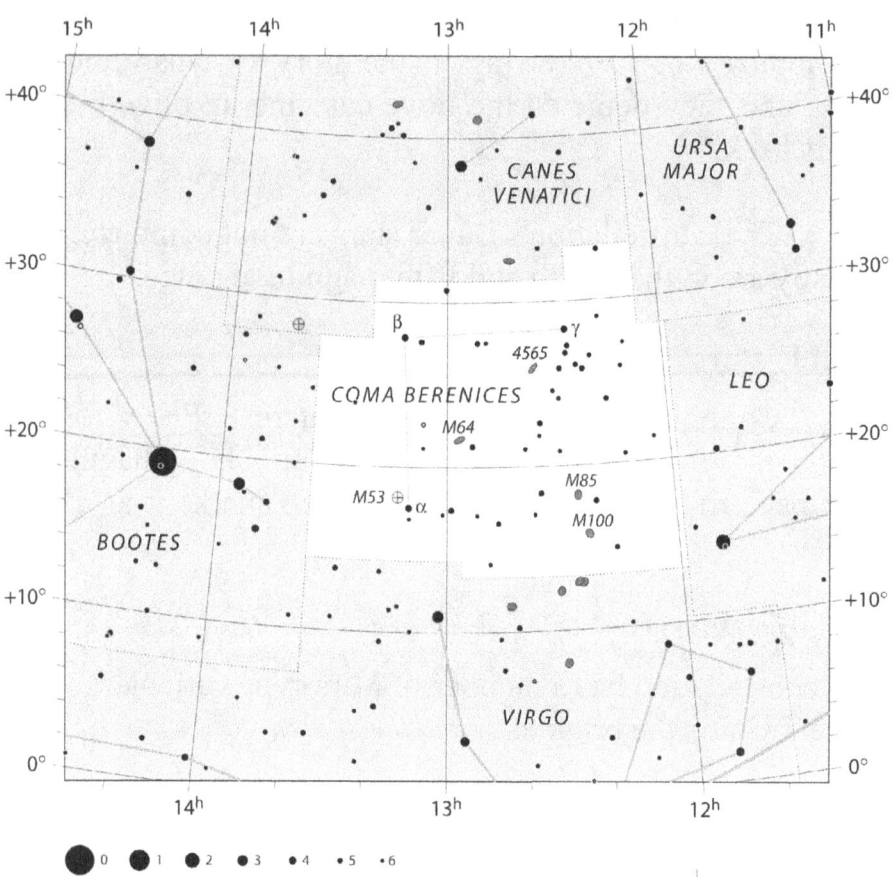

Hair of Berenices

The constellation Coma Berenices refers to a classical story concerning the hair of Berenice, the wife of Ptolemy III of Egypt. While the story is an old one, the constellation is relatively new, being introduced by Tycho Brahe (1546-1601).

According to the story, Ptolemy had waged a long war on the Assyrians, since it was they who had killed his sister. As Ptolemy returned successfully from the war, his wife Berenice had her beautiful tresses ceremoniously clipped and given to Aphrodite, laid out on the temple altar.

As the evening's festivities continued, the shorn hair was discovered to be missing. The priests might be sacrificed, if the queen's hair couldn't be found. It was the astronomer Conon of Samos who came to their rescue - proclaiming that Aphrodite had accepted the gift of Berenice's hair, which now shown brightly in the heavens next to Leo.

The stars that form the constellation really aren't that remarkable to look at, only a handful of fourth-magnitude stars, including three Bayer stars. Yet there are several fine binaries, eight Messier objects and the Coma Star cluster, not included in Messier's list.

>From *Denebola (beta Leonis)* draw a line to the bright star to the southeast, *Arcturus (alpha Bootis)*. Alpha Comae is found on this line at about the midpoint.

Now proceed north from *alpha Comae* to *beta Comae* and then west about the same distance to *gamma Comae*. These three stars form half of a nearly perfect square. They aren't very prominent, and you will have to have a nice dark night in order to study them.

Alpha Comae, sometimes called *Diadem*, has the same diameter as our Sun, and is 62 light years away with a luminosity of nearly three. It's a rapid motion binary (see below) and in the same field is the globular cluster M53 (see below).

Beta Comae is actually the brightest star in the constellation, and certainly the closest at 27 light years. It too has a diameter equal to the Sun.

Gamma Comae is an orange star about 260 light years away. It is in the same region as the well-known Coma Star Cluster, but isn't a member of that group.

Double stars:

Alpha Comae is a rapid binary of two equal stars (5.05, 5.08). The companion orbits every 25.87 years and is presently decreasing; the current (2000) separation is less than 0.05". The orbit is an unusual one, seen perfectly edge-on.

Zeta Comae is a fixed binary: (6.0, 7.5; PA 237°, separation 3.6").

17 Comae and *24 Comae* are two binaries with contrasting companions.

17 Comae is one of the members of the Coma Star Cluster. The primary is white, the companion a soft blue: 5.3, 6.6; PA 251°, separation 145.3".
From *gamma Comae* follow the slight arc of stars south that bend to the east. First comes 14 Comae, then 15, and finally 17.

24 Comae is even more spectacular: a fixed binary with an orange primary and emerald component. (5.2, 6.5; PA 271°, separation 20.3").

This binary is located eight degrees west of alpha Comae and one degree north.

35 Comae is a slow double with an orbit of over 300 years. However, unlike most long period binaries, this one is presently quite close. The companion is beginning to emerge from its close pass with the primary, gradually lengthening its separation, recently having achieved one arc second of separation. The present values are: 5.2, 7.2; PA 185° and separation 1.04".

35 Comae is in a fairly barren part of the sky, found five degrees northwest of alpha Comae.

Struve 1633 is a very pleasant fixed binary: 7.1, 7.2; PA 245°, separation 9.0". To find it start from *gamma Comae*, then drop down exactly one degree south where you'll find 14 Comae. Struve 1633 is one degree to the west.

Struve 1639 is a closer binary: 6.8, 7.8; PA 327°, 1.6". This is a slow moving binary with an orbit of 678 years.

This double star makes a small triangle with 12 Comae and 13 Comae. Start at 14 Comae and look south. The bright star to the east is 15 Comae, while below this and to the west is 13 Comae. Nearby, immediately southwest, is 12 Comae. Now look between these two stars to the southeast, where you'll find the third point in the triangle. This is Struve 1639. (Not shown on the chart due to crowding.)

Variable stars:

The constellation doesn't have a wealth of variable stars. We list the two variables that might be of some interest.

13 Comae is an alpha-CV type variable with very small range (5.15-5.18).

R Comae is a long-period variable with period of 362.82 days, and range of 7.1 to 14.6. Thus the maximums are nearly a year apart. In the year 2000 the maximum should occur in the first week of December.

Deep Sky Objects:

There are eight Messier objects (M53, M64, M85, M88, M91, M98, M99, and M100), as well as a number of other fine galaxies, with NGC 4565 being the best of the bunch.

However the best object is the unrivalled open cluster known as 'The Coma Star Cluster'.

The Coma Star Cluster:

Best seen in binoculars, the cluster fills the entire field of view: about 40 stars spread out over a five degree area.

The cluster was once known as the tuft of hair at the end of Leo's tail. It now constitutes Berenice's golden tresses.

The cluster extends south from *gamma Com* (which is not, however, a member). At about 270 light years away, the cluster is one of the closest to our solar system.

The brightest member of the cluster is 12 Comae. Other fourth-magnitude members are 13 and 14 Comae, and another thirty or so fainter stars go to make this one of the loveliest sight in the heavens.

The Messier Objects:

M53 is a globular star cluster one degree northeast of *alpha Comae*. The brightest Messier in the constellation (7.7), it tends to be most impressive with larger telescopes, which are needed to resolve the individual stars. The cluster is thought to be 65,000 light years away.

M64, the *Black Eye Galaxy*, is a bright (8.5) compact spiral one degree east-northeast of 35 Comae. The "black eye" can only be seen under ideal conditions with large telescopes. The galaxy is over 20 million light years away.

M85 is a bright spiral galaxy and member of the Virgo Galaxy Cluster, most of which is found about five degrees further south. All the remaining deep sky objects discussed also belong to this cluster.

M88 is a many-armed spiral galaxy some forty million light years away. Quite bright (9.5), it's a favourite with many Messier observers.

M91 (NGC 4548) is another spiral galaxy, but is a rather confusing object, sometimes being labelled M58. It is a rather faint galaxy (10.2) and one wonders why, with so many galaxies in the region, spreading down through Virgo, that this one was chosen by Messier.

M98 is a faint (10.1) spiral seen practically edge-on, lying just half a degree west of 6 Comae.

M99 is roughly one and a half degrees east-southeast of M98. An open spiral seen face on, its several arms are visible in large scopes. It has a brightness of 9.8.

M100 is the largest of these spiral galaxies, although difficult to appreciate in small telescopes. It's seen face-on, and has a brightness of 9.4.

Other Objects:

NGC 4565 is a well-known edge-on spiral with highly visible dust lane from end to end. It's the largest galaxy of its type and has a visual magnitude of 9.6. The galaxy is found one degree due east of 17 Comae.

Coma Berenices has many more deep sky objects, particularly the southern regions, where it borders Virgo. This is a fertile part of the sky to investigate, as the evenings grow a little warmer and more inviting.

NOTES

Corona Australis

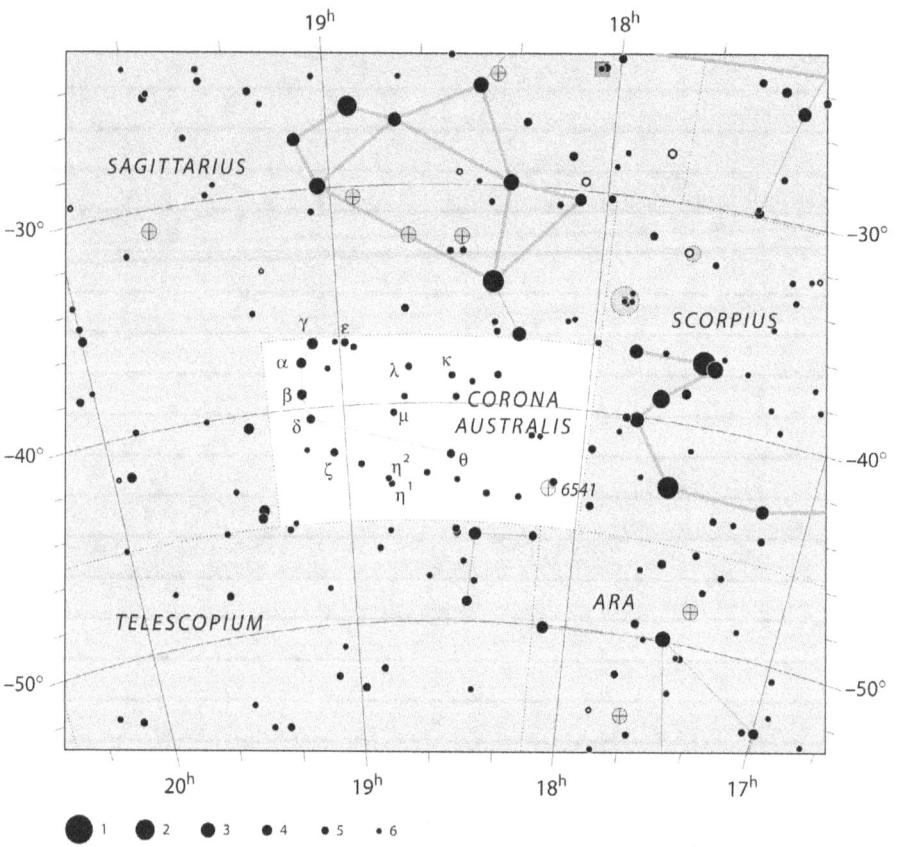

The Southern Crown

Corona Australis is a small compact constellation nestled between Sagittarius and Scorpius, just east of Scorpion's stinger.

The constellation is quite old, and is said to represent the crown worn by the centaur Sagittarius (and sometimes known as "Corona Sagittarii").

Double stars:

Kappa^2 and kappa^1 Coronae Australis form a gorgeous fixed double, visible in most of North America (as far north as Vancouver and Winnipeg) but only part of Europe, generally south of Paris or Stuttgart, and not at all in the UK.

Kappa^2 is the primary: 5.9, 6.6; 359° and separation 21.4".

h5014 is a close visual binary with an orbit of 191 years. These are two equal stars: 5.7, 5.7. Epoch 2000 values are PA 346° and separation 0.9".

Variable stars:

Corona Australis has no long-period variables, but there are several irregular variables of considerable interest. Two of these, TY and R, are found in the nebulosity NGC 6726/27/29 (see below).

Deep Sky Objects:

There are no Messier objects in Corona Australis, however the constellation does have a globular cluster suitable for binoculars, as well as an interesting region of nebulosity that goes under a multiple name.

NGC 6541 is a globular cluster, quite large and bright, about 15,000 light years away. It's nearly midway between theta CrA and theta Sco, a bit closer to the latter (more precisely thirty arc minutes east of theta Sco).

NGC 6729 is part of a nebulous region that contains both variable stars R CrA and TY CrA.

> The region is a mixture of bright and dark nebulae; the brightest region is NGC 6726/6727, which form a figure eight. Just to the SE is NGC 6729, which is much fainter, but of more interest as it contains R CrA, an irregular variable that goes from 9.7 to about 12. As the star brightens, so does the surrounding nebula.
>
> The other variable, TY CrA, is found in NGC 6726, and varies from about 8.8 to 12.6.
>
> The easiest way of finding the nebulosity is to drop seven and a half degrees south of zeta Sagittarii

NOTES

Corona Borealis

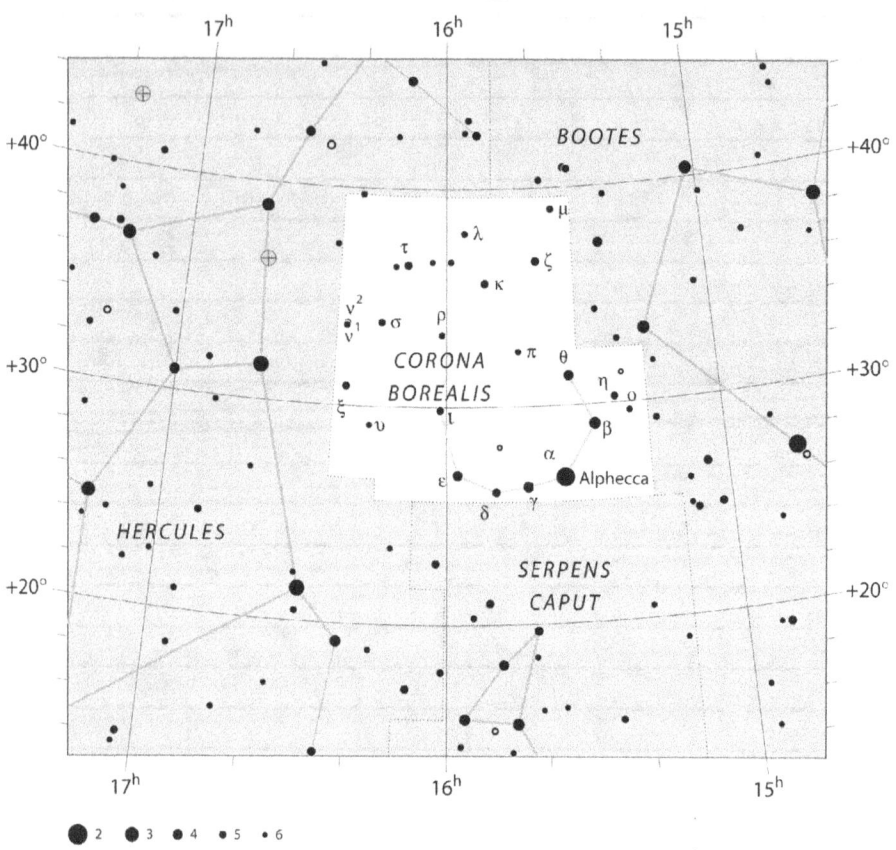

The Northern Crown

This constellation is generally associated with Ariadne, the daughter of King Minos of Crete. His wife had borne a hideous monster, half-man and half-bull, and Minos had it shut up in a labyrinth designed by the famous architect Daedalus. The maze was so complex and confusing that Daedalus "was himself scarcely able to find his way back to the entrance" (Metamorphoses VIII 166-167).

Periodically, the Minotaur needed to be fed, and a number of Athenians would be put into the labyrinth for it to eat. This happened twice; on the third feeding, the hero Theseus was one of those chosen as a sacrifice. Ariadne fell in love with him, and offered to help if he would take her away with him when he escaped. He agreed, and she gave him a thread to unwind behind him to mark his passage. He killed the Minotaur, followed the thread out of the labyrinth, and sailed from Crete with Ariadne.

The constellation Corona Borealis is found nearly midway between Arcturus and Vega; a little closer to the first of these stars. From Arcturus move up to Izar (epsilon Bootis) and then east fifteen degrees to *alpha CrB*.

The seven stars that make up the crown are not terribly bright, except for Gemma, or Alphecca (*alpha Coronae Borealis*), which is a 2.2 magnitude star 75 light years away.

The rest of the Bayer stars vary from three to six magnitude. The constellation includes several fine binaries, an unusual variable, and an extremely faint cluster of galaxies.

> Both *alpha Coronae Borealis* and *beta Coronae Borealis* are spectroscopic binaries, with periods of 17.36 days and 10.5 years respectively.

Gamma CrB is also a spectroscopic binary (period uncertain) as well as a very close visual binary (see below).

Zeta CrB is actually two stars which form a splendid binary (see below). The two are approximately 220 light years away, given their parallax of 0.015".

Double stars:

Gamma CrB (Struve 1967) is a close binary with an orbit of 91 years. The PA is 265° and separation about 0.2".

Eta CrB (Struve 1937) is a fine binary with orbit of 41.5 years. Presently the companion can be found at PA 47° and separation 0.9".

Zeta^2 and *zeta^1* *CrB* (Struve 1965): a pleasant pair of blue-white stars with 5.0 and 6.0 magnitudes; PA 305°, separation 6.3". Note that zeta^2 is the primary.

Sigma CrB (Struve 2032) is a slow binary, with a period of a thousand years. Currently the companion is at PA 236° and separation 7.03"

Nu^1 and *nu^2* *CrB* (Struve I 29) form a very wide (but only optical) pair of orange giants, quite suitable for binoculars: PA 166°, separation 372".

Variable stars:

Alpha CrB is an EA variable: 2.21 to 2.32 with period of 17.36 days.

Beta CrB is an aCV type variable: 3.65-3.72, period 18.487 days.

Gamma CrB is a delta Sct variable: 3.80-3.86, period 0.03 days (=43 minutes, 12 seconds).

Delta CrB: RS variable, 4.57-4.69.

Sigma^A CrB is an RS and delta Sct variable with period of 1.14 days.

R CrB is the most interesting variable here; an unusual RCB type variable with a range from 5.71 to 14.8.

Deep Sky Objects:

The only deep sky object is the Corona Borealis Galaxy Cluster.

This group is very faint but quite spectacular for those with the proper equipment.

The cluster is comprised of over four hundred galaxies in an area of about one degree (the width of your thumb). The galaxies are extremely distant, over a billion light years away, and consequently are very faint. The brightest of the group are 16.5 visual magnitude.

To find the cluster, move two degrees west of alpha CrB and north almost a full degree. In the same field, southwest, is the sixth magnitude binary Struve 1932 (PA 57, separation 1") with a period of 203 years.

Corvus

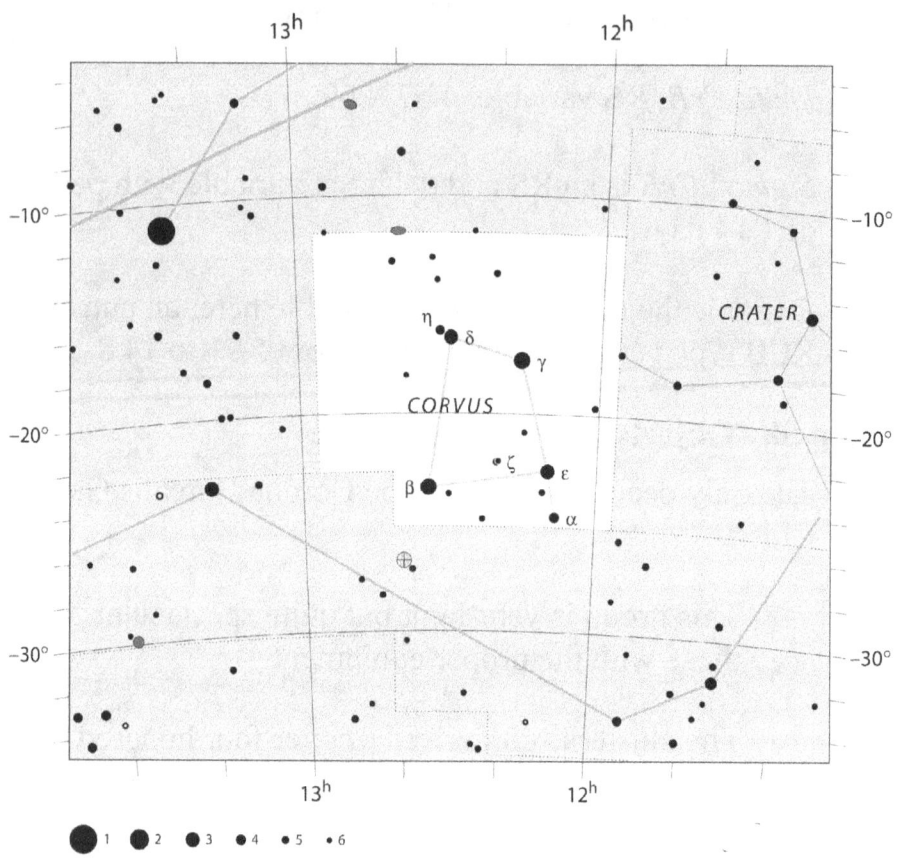

The Crow

Corvus, the Crow, was called the Raven by ancient Greeks. The story goes that Apollo sent the raven, or crow, to collect water in the nearby cup ("Crater" = goblet). But the bird wasted its time eating figs. Then, as an excuse for losing time, it gathered up the Water Snake (Hydra) in its claws and flew back, telling Apollo that this creature was the reason for its delay.

Apollo would have none of it, and threw all three: the crow, the goblet, and the water snake, into the heavens. For penance, the crow was made to suffer eternal thirst (and this makes the bird caw raucously instead of sing like normal birds).

Corvus has only a few Bayer stars. As for possible objects of interest, there are two double stars, a variable, and a curious deep sky object.

Double stars:

Zeta Corvi is an optical binary: 5.0, 13.0; PA 66°, separation 11".

Delta Corvi is a fixed binary: 3.0, 9.2; PA 214°, separation 24".

> The wide separation makes it a fine object for small telescopes; its colour contrast adds to the attraction: a white primary and purplish (or lilac) companion.

Struve 1669 is a pleasant double of equal stars: 6.0, 6.0; PA 309°, separation 5.4".

The binary is five degrees north-northeast of *delta Corvi*.

Incidentally, 1.5 degrees north of this binary is M104 (Sombrero Galaxy) in Virgo.

Variable stars:

R Corvi is a long-period variable with a range from 6.7 to 14.4 every 317.03 days. In 2000 the maximum should occur in the latter half of April.

Deep Sky Objects:

There are no Messier objects in Corvus, but there is one deep sky object of some interest: the curiously shaped Ringtale Galaxy.

NGC 4038 (the *Ringtale Galaxy*) is a rare type of galaxy classified as "peculiar". It resembles a foetus more than anything else.

Speculation has suggested it may show two galaxies in collision, or one galaxy that has broken up, split in two. However, long-exposure photography to required to bring out any detail. Burnham (pp 720-721) reproduces a number of photographs.

The galaxy is located 3.7 degrees west-southwest of *gamma Corvi*. It's about 90 million light years away.

NOTES

Crater

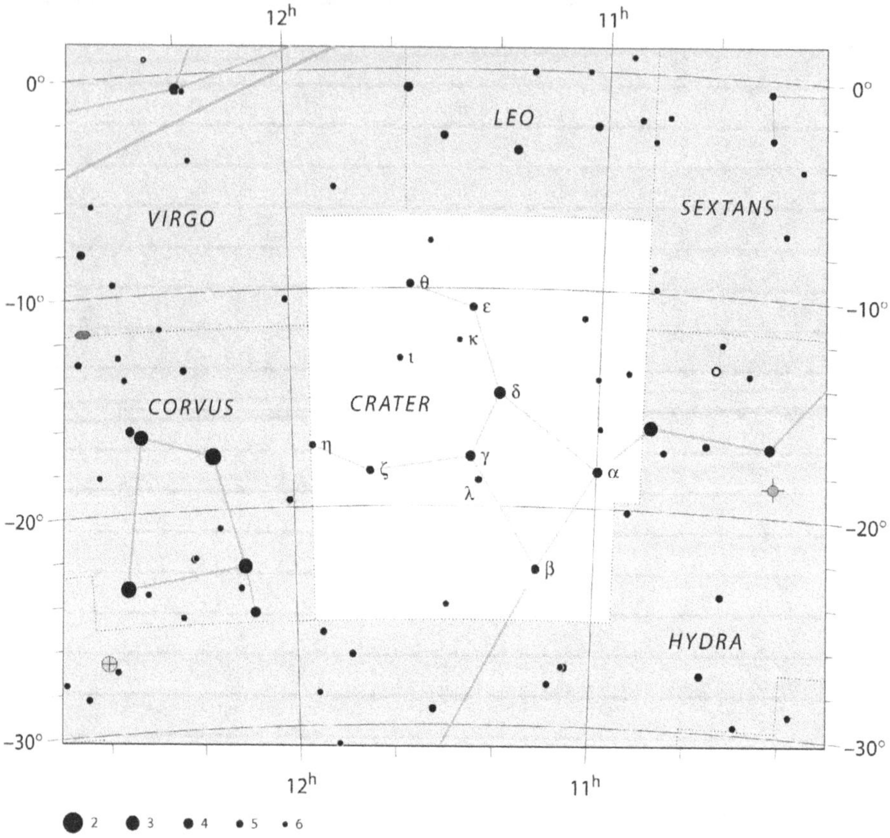

Goblet of Apollo

As told by Ovid in his Fasti, Apollo was about to make a sacrifice to Zeus and sent the crow to fetch water from a running spring. The crow flew off with a bowl in its claws until it came to a fig tree laden with unripe fruit. Ignoring its orders, the crew waited several days for the fruit to ripen, by which time Apollo had been forced to find a source of water for himself.

After eating its fill of the delicious fruit, the crow looked around for an alibi. He picked up a water-snake in his claws and returned with it to Apollo, blaming the serpent for blocking the spring. But Apollo, one of whose skills was the art of prophecy, saw through the lie and condemned the crow to a life of thirst – which is perhaps one explanation for the rasping call of the crow.

The crow is depicted pecking at the water snake's coils, as though attempting to move it so that the crow may reach the cup to drink. The cup, usually represented as a magnificent double-handed chalice of the type known in Greece as a Crater, is shown tilted towards the crow but tantalizingly just out of the thirsty bird's reach.

> *Alpha Crateris* is known as *Alkes*, Arabian for "shallow basin". The star marks the western corner of the stem, with *beta Crateris* marking the other corner.
>
> > *Alkes* is an orange (or yellow) giant, about 15 times the diameter of the Sun, and 110 light years away.

The brightest star in the constellation is *delta Crateris*, which marks the bottom of the bowl. It's about seven times the size of our Sun, and is 73 light years away.

Double stars:

> *Gamma Crateris* is a fixed binary: 4.1, 9.6; PA 96°, separation 5.2".
>
> *Iota Crateris* is a close binary: 5.5, 11; PA 226°, separation 1.4".

Psi Crateris is an even closer binary: 6.5, 7; PA 358°, separation 0.2".

Variable Stars:

None of the Bayer stars of Crater are considered to be variable. The constellation has several semi-regular variables, the brightest of which is *R Crateris*: 8.0-9.5 with a period of about 160 days.

Deep Sky Objects:

Crater has no Messier objects, and its reported deep sky objects are all very faint galaxies

NOTES

Crux

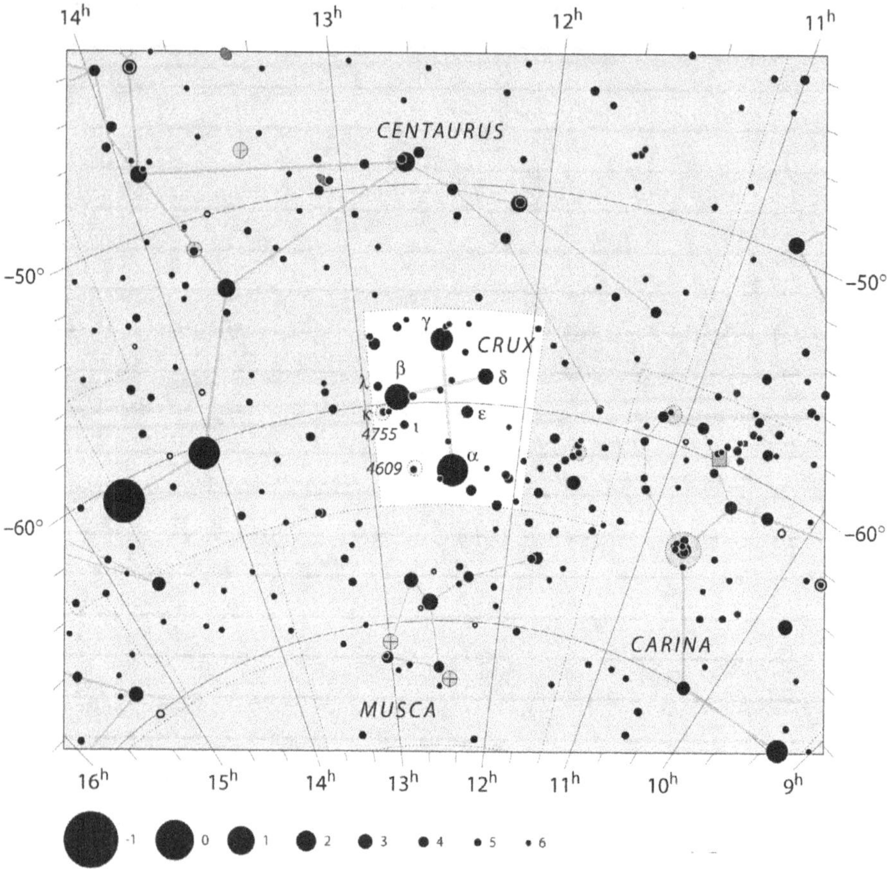

The Southern Cross

Crux first appears in its modern form on the celestial globes by the Dutch cartographers Petrus Plancius and Jodocus Hondius in 1598 and 1600 respectively; Plancius had earlier shown a stylized southern cross in a completely different part of the sky, south of Eridanus. It seems that only after he received the first accurate observations of the southern stars made by the Dutch navigator Pieter Dirkszoon Keyser did Plancius realize that the stars of Crux had been listed in Ptolemy's Almagest all along, as part of Centaurus.

The constellation's brightest star is sometimes called Acrux, a name applied by navigators from its scientific designation Alpha Crucis. It is actually a double star, divisible through small telescopes into two sparkling blue-white points. The names Becrux and Gacrux for Beta and Gamma Crucis have a similar modern origin.

Besides the cross itself, the constellation contains a unique dark nebula, a famous star cluster, and a remarkable binary.

Apart from the four bright stars that form the cross, the constellation's stars are generally fourth-magnitude. Note that while *gamma*^{A} and *gamma*^{B} are labelled as binary components, these stars only form an optical double. The two *theta* stars are also not gravitationally bound to each other; on the other hand *mu*^{1} and *mu*^{2} do form a binary system (see below).

Thousands of years ago these four stars were an object of reverence in the countries of the Near East. In the Biblical days, two thousand years ago, they were just visible at the horizon. Some might find a religious connotation, linking their disappearance with the Crucifixion of Christ. Over the millennia precession has brought the cross far to the south; it is no longer visible at latitudes north of 25 degrees.

It was the European explorers of the early sixteenth century who "rediscovered" the Southern Cross. For these adventurers the constellation was an important clock, for when it passed the meridian it was (more or less) straight up and down. Thus, by studying the constellation's inclination from the perpendicular, navigators could calculate their present time.

The principal star of note in the constellation is *Acrux (alpha Crucis)*, a splendid binary (see below). The combined visual magnitude of both stars results in a magnitude of 0.72. The stars are 320 light years away, and each is approximately one and a half to twice the size of our Sun.

> *Alpha Crucis* has an apparent proper motion of 236°. (That is, from our viewpoint, it seems to be moving very slowly in this direction.) Others in this constellation with similar motions, and therefore part of a moving star cluster, are *beta, delta, zeta, lambda,* and *mu Crucis*. The group as a whole is quite large, forming what is called the "Scorpio-Centaurus Association". See Burnham for a discussion on this cluster.

Beta Crucis (Mimosa) is the brightest star of the group, a blue-white giant (nearly five times the Sun's size) with a visual magnitude of 1.25. The star is an estimated 580 light years away, and has a luminosity of nearly 8000. The star is a variable (see below)

Gamma Crucis (Gacrux) forms the top of the cross. The reported distance may be erroneous; it's been calculated from the visual and absolute magnitudes. The resulting parallax is so large that it should be measurable.

Although gamma^A and gamma^B have been so named because of a suspected duplicity (that is, that they form a binary system) the facts are different. The stars are moving in different directions (174 degrees, 129 degrees) and are therefore not held together gravitationally.

Delta Crucis is the western arm, very similar in size and distance to *alpha Crucis*, and part of the star cluster mentioned above. The star is a beta-CMa type variable (see below).

Double stars:

Alpha Crucis is by far the best of the group: a splendid binary of equal blue-white stars: 1.58, 2.09; PA 115°, separation 4.4".

Beta Crucis has a very faint (11m) companion: PA 322°, separation 44.3".

Eta Crucis has a distance companion, rather faint: 3.6, 10; PA 299°, separation 44".

Iota Crucis is an easy binary to resolve: 4.7, 7.5; PA 22°, 26.9".

Mu^1 and *Mu^2 Crucis* form a fixed binary, also an easy one for small telescopes: 4, 5.2; PA 17° and separation 35".

Variable stars:

Crux has four *beta CMa* type variables (also called *beta Cephei* stars). These are very hot giant stars which pulsate for some inexplicable reason. Their variation is extremely small (from less than 0.01 to 0.25 magnitudes). Below are listed the *beta CMa* stars in Crux and their range.

Beta Crucis: 1.23 to 1.31 every 5h40m34s.

Delta Crucis: 2.78 to 2.84 every 3h37m30s.

Theta^2 Crucis: 4.7 to 4.74 every 2h8m1s.

Lambda Crucis: 4.62 to 4.64 every 9h28m57s.

Mu^2 is a *gamma Cas* variable, with a range from 4.99 to 5.18.

Finally, *R Crucis* isn't (as one might think) a Mira-type long-period variable, but rather a cepheid, ranging from 6.4 to 7.23 every 5d19h49m5s.

Deep Sky Objects:

"Brilliant" is the word usually used to describe *The Jewel Box* (NGC 4755). Also called the Kappa Crucis star cluster, this open cluster is composed of over a hundred stars, about fifty of which are a mixture of colourful supergiants: reds and blues intermingled with yellows and whites in a profusion of sparkling light.

The cluster is just a baby, perhaps no older than ten million years. Many of the stars have very high luminosities, approaching 100,000 Suns. The central star is *kappa Crucis*, a blue sixth-magnitude supergiant. The cluster is considered to be from 6800 to 7800 light years away.

To locate The Jewel Box, find *beta Crucis* and drop down to the southeast one and a half degrees.

The Coal Sack is a large dark nebula only 550 light years away, just to the south of the Jewel Box, visible to the naked eye.

Dark nebulae are massive clouds of interstellar gases and dust, dense enough to block out most of the light from stars behind it. The Coal Sack and Horsehead Nebulae (in Orion) are the two best known dark nebulae; of all dark nebulae, the Coal Sack is the largest one visible to the unaided eye.

NOTES

Cygnus

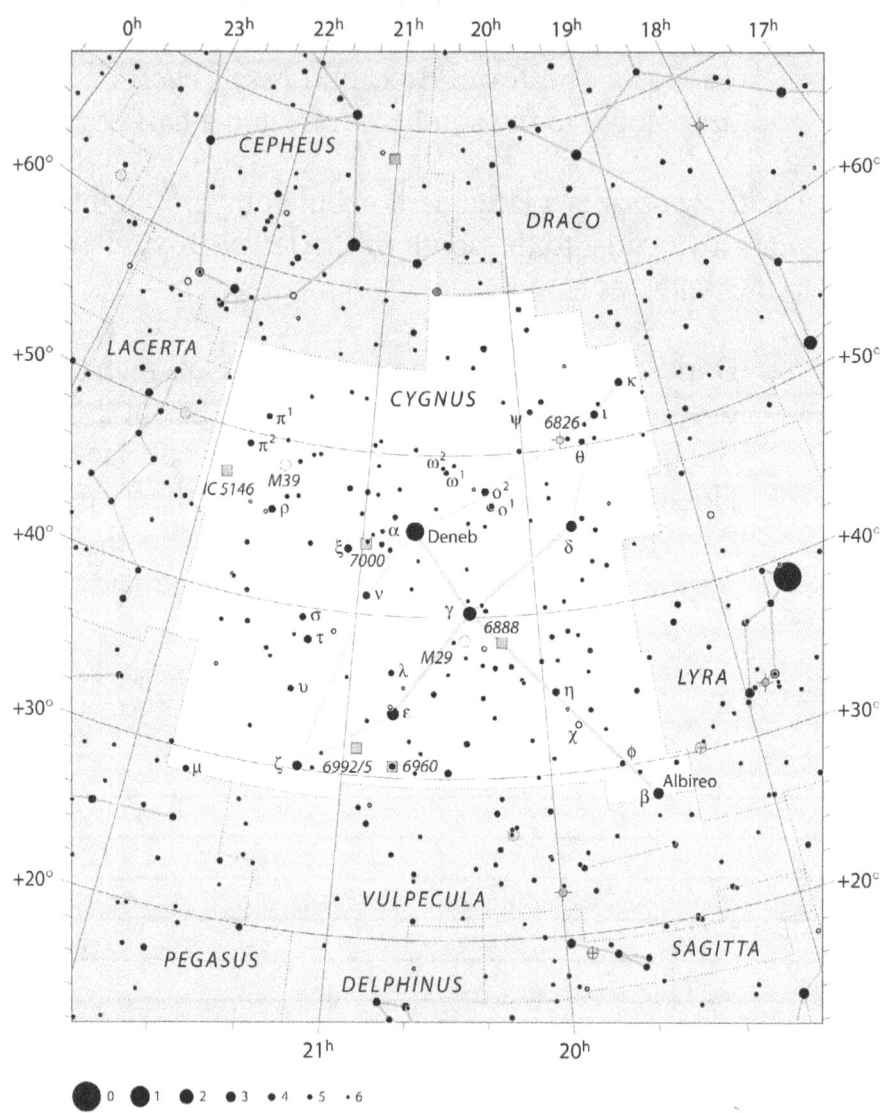

The Swan

As is the case with so many of the constellations, there are a number of possible explanations for the presence of the swan in the heavens. Some myths, for instance, state the swan was once the pet of the Queen Cassiopeia. Other versions state that the swan was Cionus, son of Poseidon, who was wrestled to the ground and smothered by Achilles. To save his son, Poseidon immortalized Cionus as a swan.

The constellation is quite bright, with the stars being generally third and fourth magnitude.

Alpha Cygni is known as *Deneb*, from *Al Dhanab al Dajajah* (the Hen's Tail). It marks the tail of the swan.

> This is a supergiant (more than a hundred times the diameter of the Sun) with a very high luminosity. Since it is so far away (3200 light years) its real brilliance is lost in space.

Beta Cygni is called *Albireo*, which is really a mistake. The words written in a sixteen-century edition of Ptolemy's *Almagest*, had been "ab ireo" (the meaning of which rests a mystery). The Arabs called it "Al Minhar al Dajajah", the Hen's Beak.

> This is a magnificent binary with a nice colour contrast (see below).

Gamma Cygni is "Sadr", from *Al Sadr al Dajajah*, "The Hen's Breast". Between gamma and beta Cygni is the Cygnus Star Cloud, a vast region of exceptional beauty.

Epsilon Cygni is "Gienah", from *Al Janah*, "The Wing".

The constellation has several superb visual binaries as well as one of the more intriguing Mira-type variables. Several faint deep sky objects are also found in Cygnus, but it seems surprising that, while the constellation lies in the heart of the Milky Way, it has no truly outstanding clusters, nebulae, or galaxies.

Double stars:

Beta^1 and beta^2 form an extraordinary binary: gold and blue (or perhaps yellow and blue-green).

> The component is quite wide, making it a popular object for binoculars.

> AB: 3.1, 5.1; PA 54 degrees, separation 34.3".

Delta Cygni is a visual binary with an orbit of 828 years. Presently the values are: 2.9, 6.3; 224°, 2.5".

Mu Cygni is another visual binary (4.8, 6.1) with a long orbit, 789 years. For the next fifty years the orbit will continue to appear to approach the primary (as seen from the earth). The 2000.0 values are: 309°, 1.85".

Tau Cygni is a visual binary with a 49.9 year orbit: 3.9, 6.8. The 2000.0 year values are PA 328°, separation 0.8".

30 Cygni and *31 Cygni* [omicron^1] form a wonderful triple, suitable for binoculars:

> **AB: 4.0, 5.0; 333° and separation 338"**
> **(orange and turquoise).**
> **C: 7.0; 173°, separation 107" (blue).**

61 Cygni is another fine binary of two orange stars: 5.2, 6.0. The 2000.0 values are PA 150°, and separation 30.3".

61 Cygni also holds the distinction of being the first star to have its parallax measured. This occurred in 1838, by Friedrich Wilhelm Bessel, a German astronomer.

Variable stars:

Cygnus has many variable stars, most of which are too slight to notice without high-tech equipment.

Alpha Cygni is the prototype for a variable class of pulsating supergiants. These variables have a spectral type of A or B and very high absolute magnitudes.

> Some fifteen supergiants are members of this group (including kappa Cas). The period ranges from five to ten days and the amplitude is less than 0.1 magnitude. For alpha Cygni, the range is 1.21 to 1.29.

Tau Cygni is a delta Scuti type variable, ranging from 3.65 to 3.75.

Upsilon Cygni is a gamma Cas type variable: 4.28-4.50.

Chi Cygni is by far the most interesting variable of the constellation. This is a Mira-type variable with period of 408.05 days. It takes several months to reach its maximum, then several more before it disappears from sight, at a minimum of 14.2.

> The maximum varies, mostly it winds up in the 4.3-4.5 range, although it has been known to achieve third-magnitude status. Burnham has a finder's chart, but if it

is in the fifth or fourth magnitude range, you should have no difficulty in finding it: it has a bright red colour, and is located about two degrees SW of eta Cygni (or about one quarter the distance from eta to beta Cygni).

The star's next appearance should be in mid September of 1997. In 1998 it should reach its maximum in the last week of October.

Deep Sky Objects:

Cygnus contains two rather uninteresting Messier objects and some faint and difficult nebulae:

> *M29 (NGC 6913)* is an open cluster, quite a lackluster Messier of about half a dozen eighth magnitude stars shaped like a square. The cluster is found 1.5 degrees south of gamma Cygni (and a couple of arc mintues to the east).

> *M39 (NGC 7092)* is large and scattered and equally unspectacular; a group of faint (seventh magnitude) stars forming a rough triangle. It is nine degrees ENE of alpha Cygni.

> *NCC 7000* is called "North American Nebula" because of its shape. It's a bright slightly greenish emission nebula.

>> The nebula is described in most references as "bright" but in fact you'll find it is extremely faint. It is best seen in binoculars, and is found between alpha Cygni and xi Cygni.

The Veil Nebula West (NGC 6960) and *The Veil Nebula East (NGC 6992/95)* are fine filaments seemingly suspended in the cosmos. It takes quite a large scope, perfect conditions, and plenty of patience to appreciate their delicate lines.

> The nebulae are 2.5 to 3 degrees south of epsilon Cygni. The star 52 Cygni is in the same field as the western segment, and is the best starting point to search for the elusive nebulae.

> *52 Cygni* is three degrees due south of epsilon Cygni (and it's a binary as well, *Struve 2726*: 4,9; 67°, 6.6").

Cygnus A is the second brightest source in the 'radio sky', after the supernova remnant known as Cas A. This pecularly-shaped galaxy is considered to be a billion light years distant, and is an object of intense investigation. Two lobes of radio emission are fed by jets of energetic particles from the galaxy core. (I thank Philip Blanco for this description. Philip has a web page devoted to Cygnus A. Those interested just set their search engine to 'Cygnus A'.)

> *Cygnus A* is found in a highly nebulous region of the constellation, about three and a half degrees west of gamma Cyg.

Delphinus

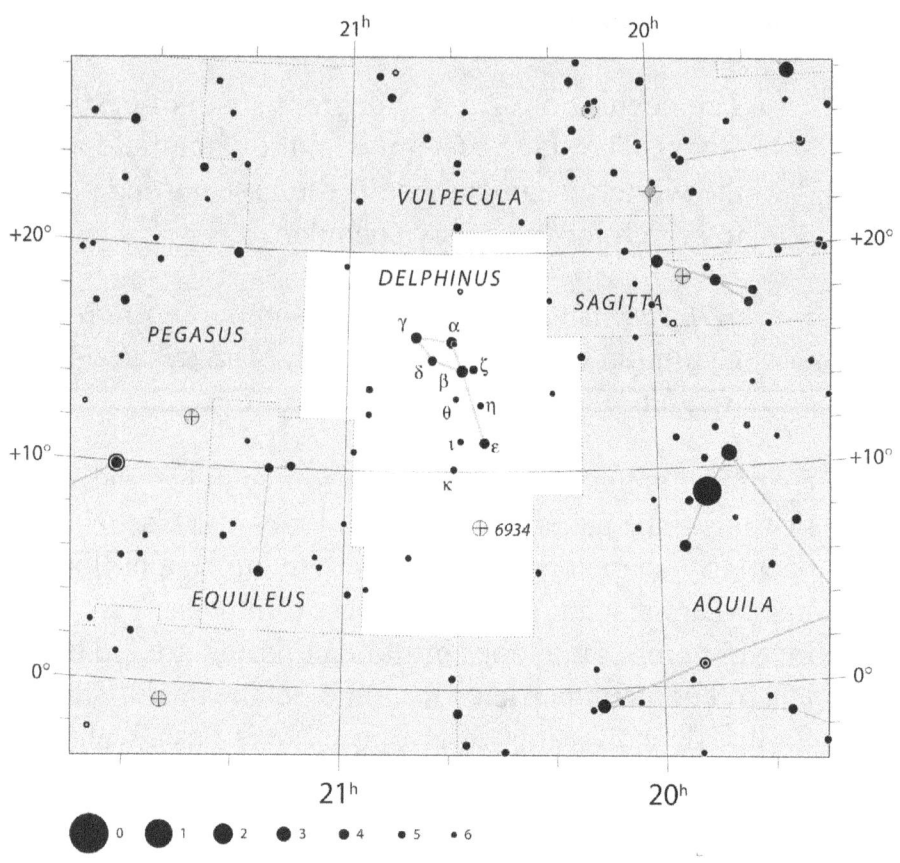

The Dolphin

Delphinus, "The Dolphin", is an ancient constellation located just west of Pegasus.

Some references name a certain "Arion" as the inspiration for the constellation.

There were two Arions in antiquity. One was a (mythic?) poet who may have lived in the eighth century BC. This Arion, traveling from Sicily to Corinth, was thrown overboard by the ship's crew, eager for the valuables he was carrying. A dolphin is said to have rescued the poet. But this dolphin probably isn't the constellation's origin.

The second Arion was a son of Poseidon and Demeter, and was in fact a horse (like his half-brother Pegasus). Instead of hooves, he had feet on his right side. And, unlike most horses, he could talk. But this Arion also has nothing to do with the constellation.

It is most likely however that the constellation is associated with Poseidon. It was probably his way of thanking one of his messengers for a job well done.

As God of the Sea, Poseidon had fifty sea-nymphs at his court. These were all born of Nereus and known therefore as the Nereids.

While Poseidon had many casual love affairs, when he set out to find a wife he was concerned that she be accustomed to life in the sea. His first choice was Thetis, one of the fifty Nereids. But he learned that any son born of Thetis would grow to become greater than his father. Clearly Poseidon couldn't accept that prophecy.

As a side note, Thetis married Peleus, a mortal,

and they had a famous son named Achilles. Thetis dipped Achilles in the river Styx to make him invulnerable to his enemies. As most people now knows, since his mother grasped him by the heels, they were the only part of Achilles which were vulnerable. Wouldn't you know, the day would come when he'd get a poisoned arrow in his heel and die from it.

Poseidon's next choice in marriage was a sister of Thetis, called Amphitrite. But when Poseidon pressed Amphitrite to marry him, she was quite disgusted by the thought and fled to the far-off Atlas Mountains. Poseidon sent a number of messengers to persuade her to return, as his wife, to his underwater realm.

The messenger who succeeded in this task was the dolphin Delphinus. Amphitrite was so beguiled by Delphinus' pleadings she relented and returned to Poseidon and became the Queen of the Sea. They had many children.

Delphinus was later put in the heavens as a constellation by a grateful Poseidon.

The asterism is rather curious, for its four main stars form a rectangle called "Job's Coffin". This is probably a hang-over from the time Delphinus was interpreted as a whale, as in Chapter 41 of Job where God challenged Job: "Canst thou draw out leviathan with an hook?" However there is no reference to Job being swallowed by a whale, as happened with Jonah, so the name Job's Coffin remains a bit of a mystery.

Double stars:

Delphinus has several fine binaries, a Mira-type variable, and a very remote globular cluster.

Beta Delphini is a very close visual binary with orbit of 26.7 years. Epoch 2000 values: 4.0, 4.9; PA 343°, separation 0.5".

Gamma^1 and *gamma^2 Del* form a fine binary with (perhaps) subtle color change (observers argue over this; some find them both yellow, others that the companion is greenish or bluish): 4.5, 5.5; PA 268°, 9.6"

Struve 2725 is a wonderful sight in the same field as gamma Del (to the SW): 7.3, 8.0; PA 9°, separation 5.7".

Variable stars:

R Delphini is a Mira-type variable with a period of 285.07 days and a range of 7.6-13.8. In the year 2000 the maximum should occur near the end of August.

Deep Sky Objects:

NGC 7006 is a very remote globular cluster, perhaps as far as 200,000 light years away. Because of its distance it is extremely difficult to resolve. It is located fifteen arc minutes due east of gamma Delphini.

NOTES

Dorado

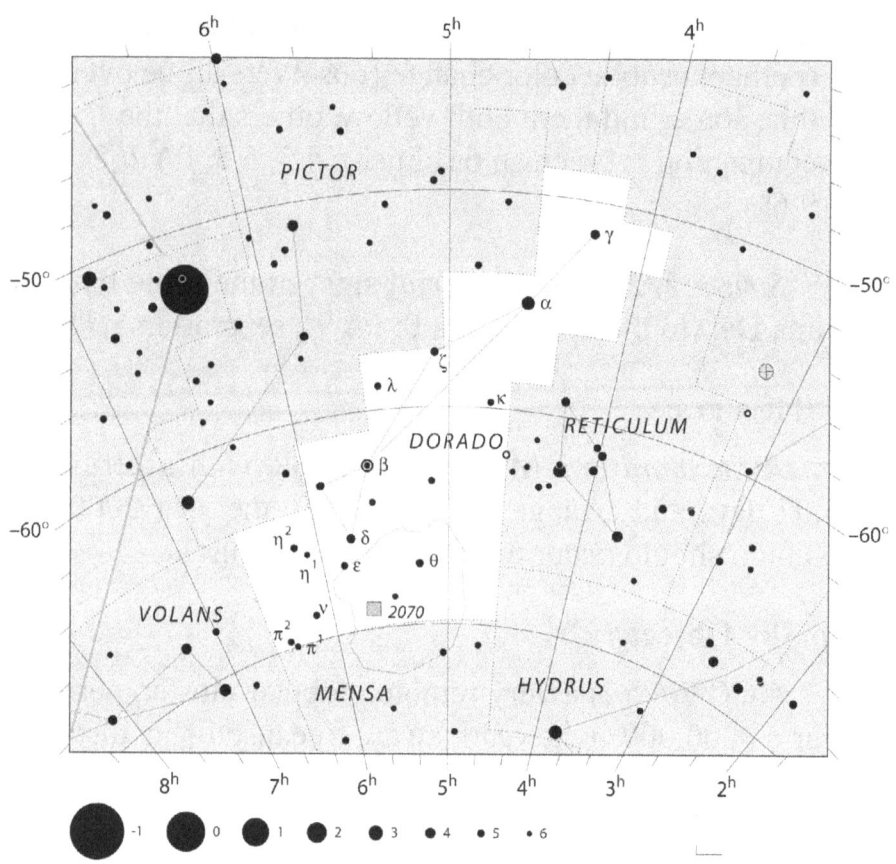

The Goldfish

𝔇orado was introduced by Johann Bayer in 1603 in his epoch-making star atlas, Uranometria.

Dorado, "The Goldfish", is also known as "The Swordfish". While the Bayer stars are not very bright, there are several objects of interest in the constellation, notably the Large Magellanic Cloud and the Tarantula Nebula.

Double stars:

While Dorado has its share of binaries, they are quite dim and not particularly interesting. The one exception might be *h3796*.

h3796 is a multiple group found in the vicinity of NGC 2070 (see below for this deep sky object).

The principal star is nine-magnitude while most of the other dozen or so companions are in the 12-14 magnitude category.

AB: 9, 10; PA 213°, separation 0.5".

A full listing of the other companions' position angles and separations is given in Burnham (p.833).

Variable stars:

Alpha Doradus is an alpha-CV type variable: 3.26-3.3 every 2d 23m.

Beta Doradus is a cepheid: 3.46-4.08 every 9d20m.

Gamma Doradus is an EW type variable: 4.23-4.27.

R Doradus is a semi-regular variable, 4.8 to 6.6 about every 338 days.

Deep Sky Objects:

NGC 2070, the Tarantula Nebula, is a gaseous section of the Large Magellanic Cloud. The nebula is so bright it goes by the name *30 Doradus*. Dozens of supergiant stars are clustered at its centre, furnishing the nebula's light. With a diameter of about a thousand light years, if the Tarantula Nebula were moved to Orion - at the same distance as the Orion Nebula - then it would entirely fill the constellation of Orion.

The Large Magellanic Cloud is a miniature galaxy about 200,000 light years away, a satellite of the Milky Way. It has perhaps a tenth of the mass of our own Milky Way Galaxy, with roughly 10,000 million stars (or ten billion if you wish).

NOTES

Draco

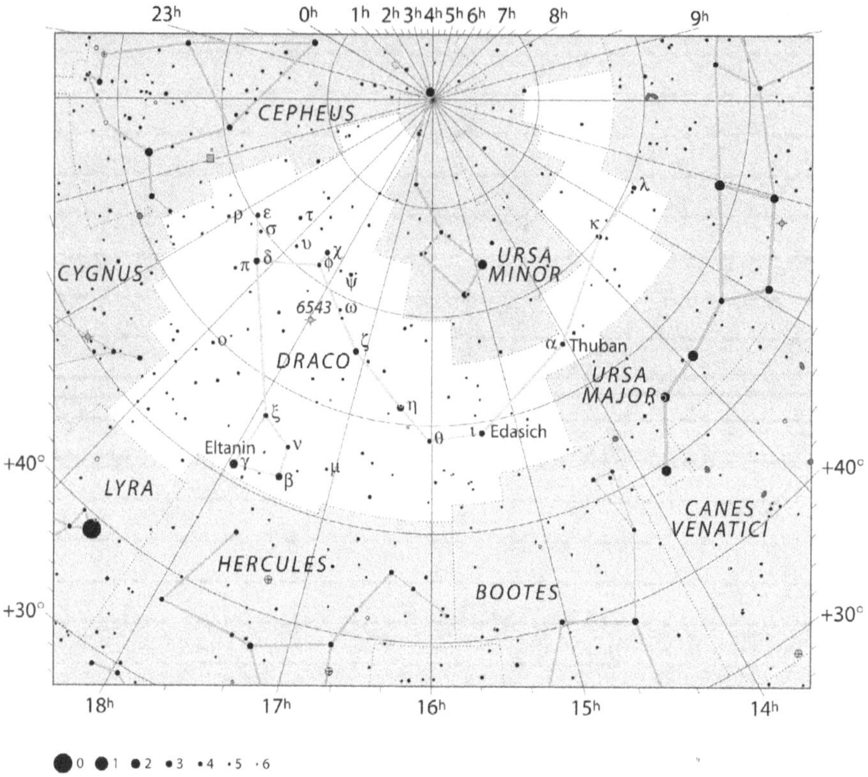

The Dragon

It is unclear precisely which mythological dragon Draco represents. There are, however, three main contenders.

One version--the least likely--of the Draco story is that the dragon fought Minerva during the wars between the giants and the gods. Minerva threw Draco's twisted body into the heavens before it had time to unwind itself.

Another possibility is that Draco represents the dragon who guarded the golden apples in the garden of the Hesperides. One of the labors of Heracles was to steal these apples (some sources state it was his eleventh labor, others it was his twelfth). This was, according to Bulfinch, the most difficult labor of all..., for Heracles did not know where to find them. These were the apples which Hera had received at her wedding from the goddess of the Earth, and which she had entrusted to the keeping of the daughters of Hesperus, assisted by a watchful dragon. After various adventures, Heracles arrived at Mount Atlas in Africa. Atlas was one of the Titans who had warred against the gods, and after they were subdued, Atlas was condemned to bear on his shoulders the weight of the heavens. He was the father of the Hesperides, and Heracles thought might, if any one could, find the apples and bring them to him (Bulfinch's Mythology, 136).

Heracles suggested this plan to Atlas, who pointed out two problems: first, he could not simply drop his burden; second, there was the awful guardian dragon. Heracles responded by throwing his spear into the garden of the Hesperides and killing the hundred-headed beast, and then taking the burden on his own shoulders. Atlas retrieved the apples and, reluctantly taking the burden onto

his shoulders once again, gave them to Heracles. Hera placed the dragon
in the heavens as a reward for his faithful service.

By far the most commonly accepted version of Draco's arrival in the heavens, however, is that Draco was the dragon killed by Cadmus. Cadmus was the brother of Europa, who was carried off to Crete by Zeus in the form of a bull (Taurus). Cadmus was ordered by his father to go in search of his sister, and told he could not return unless he brought Europa back with him. "Cadmus wandered over the whole world: for who can lay hands on what Jove has stolen away? Driven to avoid his native country and his father's wrath, he made a pilgrimage to Apollo's oracle, and begged him to say what land he should dwell in" (Metamorphoses III 9-11).

Cadmus followed Apollo's advice and found a suitable site for his new city. He sent his attendants to find fresh water to offer as a libation to Zeus, and they wandered into a cave with springs. As they were getting water, however, they were all killed by "the serpent of Mars, a creature with a wonderful golden crest; fire flashed from its eyes, its body was all puffed up from poison, and from its mouth, set with a triple row of teeth, flickered a three-forked tongue" (Metamorphoses III 31-34). After his companions did not return, Cadmus himself went into the cave and discovered the dragon. He killed it with his spear, and then (upon the order of Minerva) sowed the dragon's teeth in the ground. From the teeth sprung warriors, who battled each other until only five were left. These five, along with Cadmus himself, were the first people of the city of Thebes.

It is interesting, however, to note that Ovid himself

does not equate the dragon of Mars with Draco. In fact, in book III of Metamorphoses, he describes the dragon killed by Cadmus in terms of the constellation: "It was as huge as the Serpent that twines between the two Bears in the sky, if its full length were seen uncoiled" (45-47).

The Serpent described by Ovid is certainly the same one as we see today, twisting past Cepheus and between Ursa Major and Ursa Minor in the north, with its head beneath the foot of Heracles. Its location therefore seems to fit best with the myth that Draco was the dragon in the garden of the Hesperides.

it wraps itself around the northern hemisphere Draco is circumpolar, not far from the North Pole. In fact Thuban (*alpha Draconis*) was once the Pole Star, at about the time these stories were being told for the first time.

A very old and extensive constellation, Draco once held even more stars. Quite fittingly, Hercules is just to the east of Draco. In fact, some cartographers draw the figure of Hercules with one foot resting on the head of Draco.

Depending on the time of year one studies the constellation, its head (formed by *beta, gamma, nu,* and *xi*) takes on a different look. When *beta* and *gamma* are `on top', they look like two eyes, or perhaps the forehead. Other times of the year the face is rather indistinct.

There are a full range of Bayer stars in the constellation. While there are few deep sky objects of any interest (and just one Messier) the constellation does have a wide variety of interesting binaries to investigate, some of which are listed further below.

Thuban is the Arabic name for Dragon. To find Thuban sweep down the length of the Little Dipper and jump over to the end of the handle of the Big Dipper. Midway is found a much fainter star, which is Thuban. It is believed that the star was considerably brighter several thousand years ago.

> This star was the pole star at about 2700 BC. The fact that Thuban was the Pole Star at just the time the Egyptians were building pyramids hasn't escaped the archaeologists.
>
> The main object of the archaeologists' study is the Great Pyramid of Khufu. It is claimed that a particular passage in the pyramid was built to point at Thuban as that star dipped to its lower culmination.
>
> However, if the above assumption is true, then the pyramid would have been built at around 2200 BC. The problem is that Khufu is about five hundred years older.
>
> There are many books and articles on the subject (and no doubt several web pages on the Internet) for those who wish to delve deeper into the problem or to study the alignment of other stars with ancient artifacts.

Double stars:

Draco has dozens of binaries worth investigating. Below are some of the more easily resolved double star systems, and a couple not-so-easy ones.

> *Mu Draconis* is one of the closer binaries, a slow orbit of 482 years. Presently the companion can be found at

PA 14° and separation 1.91".

Nu Draconis is a splendid fixed binary, found in the dragon's head. Two similar 4.9 visual magnitude stars: PA 312° and separation 61.6".

Psi Draconis is also easily resolved: 4.9, 6.1; PA 15°, separation 30.3"

Omicron Draconis has a fine colour contrast, orange and blue. Magnitudes 4.7, 7.5; PA 326°, separation 34.2".

17 Draconis forms a magnificent fixed triple with 16 Draconis. 17AB: 5.5, 6.4, PA 108°, separation 3.4"; 16 Draconis is component C: PA 194, separation 90.3".

26 Draconis is a close binary with orbit of 76 years. The component is currently at PA 334° and separation 1.6". There is a faint (10m) very wide third member, at PA 162° and separation 12.3'.

41 and 40 Draconis (Struve 2308) form a pleasant, fairly wide, binary of two cream-coloured stars: 5.7, 6.0: PA 232°, 19.3". Note that 41 is the primary.

Struve 2398 is an extremely near binary at only 11.3 light years. It consists of two red dwarfs, 8.0, 8.5; PA 163°, separation 15.3". It is thought the companion has an orbit of roughly 350 years.

> The binary is found just between omicron Draconis (which to the east) and 39 Draconis.

Variable stars:

R Draconis is a Mira-type variable with a period of 245.6 days; it fluctuates from 6.7 to 13.2 magnitude. In 2000 the maximum should occur in the third week of April.

Deep Sky Objects:

Draco offers one Messier object: M102 (although this object is not universally recognised as a bona fide Messier). With several dozen other galaxies, and a bright planetary nebula, there are plenty of objects to study. Below are a few suggestions.

M102 (NGC 5866) is an edge-on galaxy with dust lane and brightly glowing centre.

The galaxy is four degrees southwest of *iota Draconis*.

NGC 5907 is in the same region one degree east of M102. This is another edge-on (nearly flat) galaxy with dust lane.

NGC 5985 is an inclined spiral, quite faint unless under ideal conditions.

NGC 5985 is midway between *iota* and *theta Draconis*; (NGC 5982 is in the same field to the west. This elliptical gallaxy is considerably smaller but about the same magnitude, around 12).

NGC 6543: a planetary nebula that appears as a miniscule blue-green disk. Because of its blue-green colouring, it is sometimes called the Cat's

Eye Nebula. It's located halfway between *delta* and *zeta Draconis*. It's exact distance isn't known; estimates vary from 1500 to 3500 light years.

NOTES

Equuleus

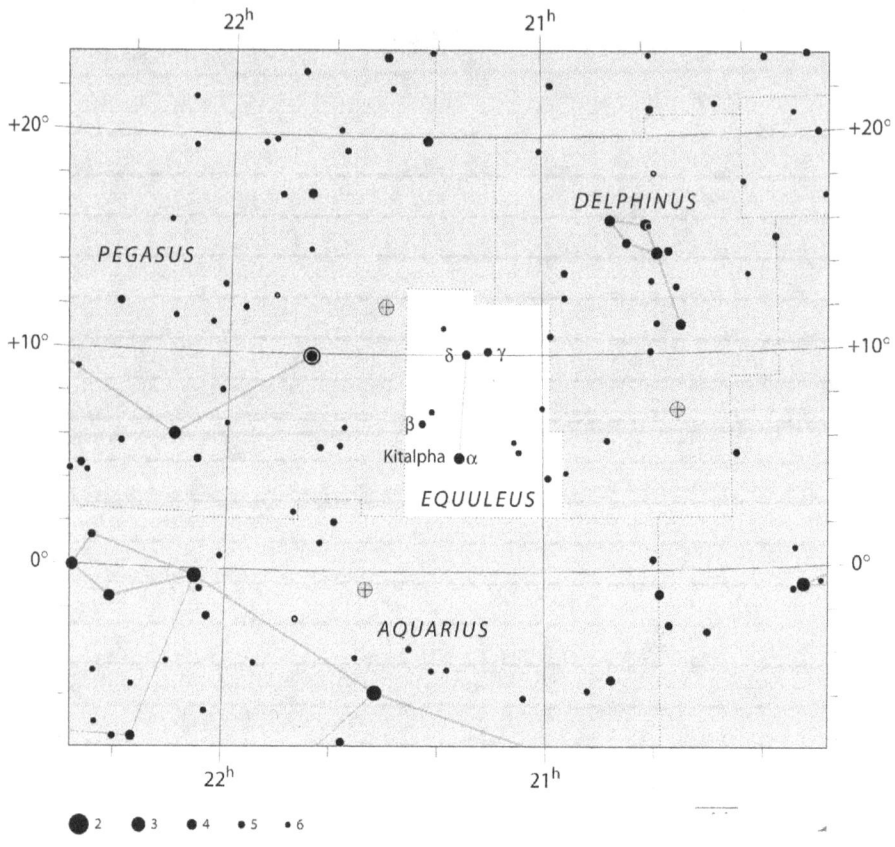

The Little Horse

Equuleus, "The Little Horse", is one of the smallest constellations in the heavens. It's quite old, and may have been founded by Ptolemy in the second century AD. However the author of the Almagest often borrowed from others and it is possible his principal source, Hipparchus, was the true creator of this constellation.

The outstanding Greek astronomer Hipparchus (fl46-127 BC) composed the first star catalog, of about 850 stars. He also discovered the precession of the equinoxes and invented trigonometry. It is not known if he actually created any constellations.

The "little horse" that the name refers to is lost in antiquity. Some sources believe it to be a half-brother of Pegasus, Celeris. However I've not found any reference to this character. The only brother of Pegasus that I've come across is Chrysaor, born simultaneously with Pegasus. Instead of a horse, Chrysaor was a warrior.

Its original name seems to have been *Al Faras al Awwal* and the Latin equivalent *Equus Primus*, "the First Horse", since it rises just before Pegasus.

The asterism is a nondescript triangular form made from the brightest four stars. The brightest of these, alpha Equulei, is called *Kitalpha*, from *Al Kit ah al Faras*: "Part of the Horse".

There are only a half dozen Bayer stars, which are generally fifth magnitude.

There are a number of multiple binaries here, but little else.

Double stars:

Gamma Equulei (also known as 5 Equ) is a multiple system with quite faint components (although C is only optical).

AB: 4.7, 11; PA 268°, separation 1.9".
C: 12; PA 5°, 47.7".
D: 6; PA 153°, 352".

Delta Equulei (also known as Struve 2777) is a multiple system including one of the most rapid visual binaries, with an orbit of only 5.7 years: 5.2, 5.3; 2000 values: PA 33°, 0.2". AC: 9.4, PA 14°, 47.7".

Epsilon Equulei (Struve 2737) is also a multiple system:

AB: 5.2, 6.0; PA 288°, 0.9"
C: 7; PA 70°, 10.6".
D: 12.5; PA 280°, 74.8".

Lambda Equulei (Struve 2742) is perhaps the most attractive binary in Equuleus. It has two equal but rather faint stars:

7.4, 7.4; PA 218°, separation 2.8".

Variable stars:

None of Equuleus' variables are suitable for amateur viewing; the brightest Mira-type variable (R Equ) only gets to a visual magnitude of 8.7 every 261 days.

NOTES

Eridanus

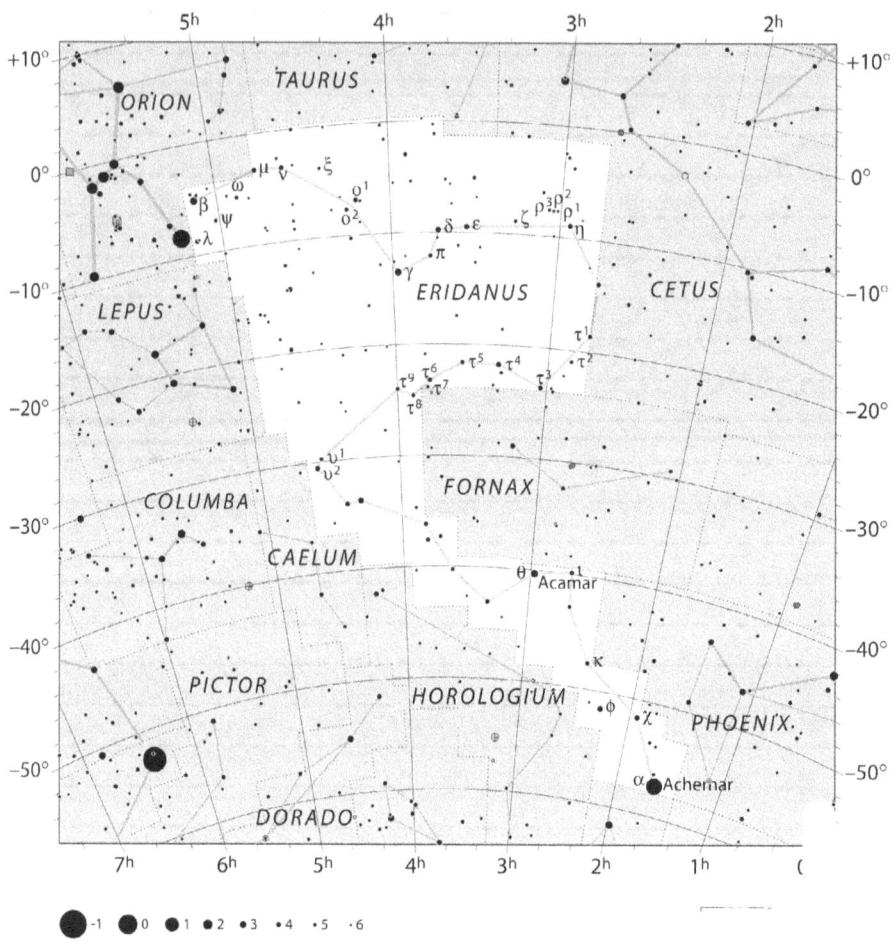

The River

Eridanus is a river in northern Italy, now known as the River Po. Called by Virgil the "king of rivers," Eridanus was made famous in connection with the death of Phaethon.

Phaethon was the son of Phoebus Apollo and the nymph Clymene. For his birthday one year, Phaethon asked his father for some proof that he was indeed the son of the sun-god. Apollo said he would give the boy any gift he desired as a token of his fatherly love, and Phaethon promptly asked for the chance to drive the chariot of the sun. His father balked, knowing that no mortal youth could possibly have the strength necessary to control the horses. However, Phaethon insisted, and Apollo had granted his word.

Eridanus' stars are fairly faint, except for alpha (Achernar, "the end of the river"). This bright star is only visible in latitudes south of 33 degrees, which eliminates most of North America and all of Europe.

Curiously *theta Eridani* bears nearly the same name, Acamar. This is a variation on Achernar since this star was once the river's terminus and it then bore the name now carried by *alpha Eridani*.

Double stars:

Theta^1 and theta^2 Eridani form an attractive pair: 3.4, 4.5; PA 88°, separation 8.2".

Omicron^2 Eridani is an interesting triple system only 15.9 light years away.

AB form a very wide pair, with orbit of perhaps 8000-9000 years: 4.5, 9.7; PA 105°,

separation 82.8".

BC form a close visual binary with orbit of 252 years. The primary is a white dwarf, with about twice the diameter of the earth. The component a red dwarf.

In fact omicron^B is considered the easiest white dwarf for amateur telescopes. The component C has an extremely small mass, considered to be about 0.2 solar mass.

9.7, 10.8; presently PA 337° and 9.3" separation.

32 Eridani is another attractive pair, with colour contrast - yellow and blue: 5.0, 6.3; 347°, 6.9" separation.

32 Eri is near the northern boundary, 10° west of nu Eri (or about 10° north of gamma Eri).

p Eridani is a visual binary near the southern boundary of the constellation. It has an orbit of 483.7 years: 5.8, 5.8; currently the component is at PA 191° and separation 11.5".

The binary is found one degree north of Achernar (alpha Eri).

Variable stars:

Eridanus has a variety of variables, most of which are too small magnitude for amateur observation. Below are only a few examples.

Gamma Eri is an Lb type variable: 2.88-2.96.

Delta Eri is an RS CVn type variable: 3.51-3.56.

Lambda Eri is a beta Cep variable: 4.22-4.34, it has a period of 16h 50m.

Nu Eri is also a beta Cep variable: 3.4-3.6 with period of 4h 16m.

Deep Sky Objects:

Eridanus has many galaxies, but most are quite faint. Below are two of the best examples, both found in the vicinity of tau^4 Eri.

NGC 1232 is a spiral galaxy seen face on. It's 2° NW of tau^4 Eri.

NGC 1300 is a splendid barred spiral. It's 2.3° due north of tau^4 Eri.

NOTES

Fornax

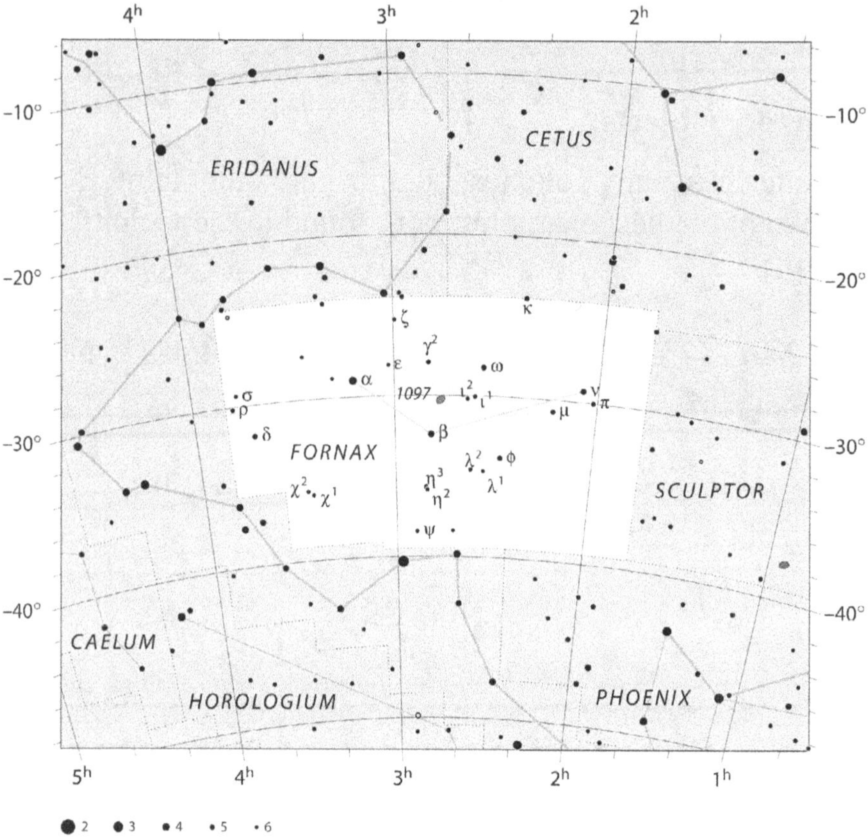

The Furnace

Fornax, "The Furnace", is another of those constellations created by Nicolas Louis de Lacaille in the mid-eighteenth century; he took several dozen fairly bright stars away from the middle of Eridanus and called the result Fornax Chemica.

Alpha Fornacis is an easy visual binary, however the most noteworthy object in this constellation is the Fornax Galaxy Cluster (see below).

The asterism is a rather uninspired connection of its three brightest stars. In general the Bayer stars are fourth and fifth magnitude.

Double stars:

Alpha Fornacis is a visual binary with orbit of 314 years. The 2000 epoch values are 4.0, 7.0; PA 299°, separation 5.1".

Gamma^{1A} and *gamma^{1B} For* form a faint triple system:
.

AB: 6.0, 12; PA 145°, 12".
C: 11; 143°, 41".

Omega For is a fixed binary: 5.0, 7.7; PA 244°, 10.8".

Variable stars:

Nu Fornacis is an alpha CV type variable: 4.68-4.73 every 1.9 days.

Deep Sky Objects:

In the southeastern corner of Fornax, on the border with

Eridanus, is a collection of bright galaxies which go by the collective name of "The Fornax Galaxy Cluster".

Eighteen galaxies make up the cluster. The brightest is *NGC 1316* (which is also a radio source called *Fornax A*). *NGC 1399* is nearly as bright; the third brightest, *NGC 1365*, is a splendid barred spiral seen face on, with open spiral arms.

NGC 1316 is 2° SW of chi^3 For;
NGC 1365 is 1° ESE of chi^3 For;
NGC 1399 is 2° E of chi^2 For.

NOTES

Gemini

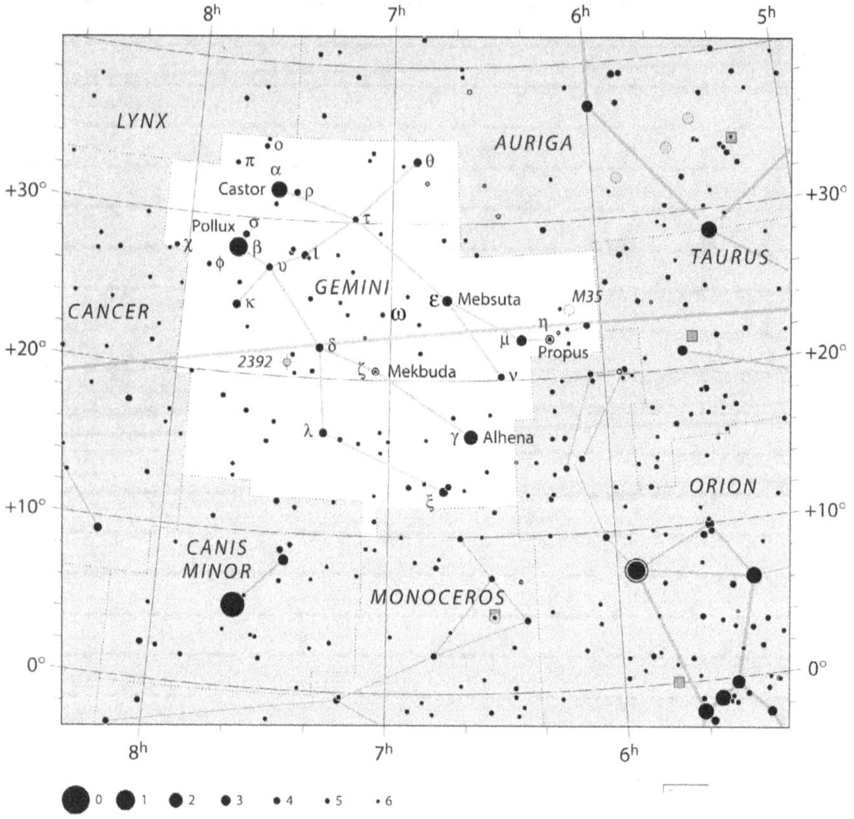

The Twins

Gemini is a zodiacal constellation representing the twin brothers Castor and Pollux. Both were mothered by Leda, and were therefore brothers of Helen, but they had different fathers: In one night, Leda was made pregnant both by Zeus in the form of a swan and by her husband, the king Tyndarus of Sparta. Pollux, as the son of a god, was immortal and was renowned for his strength, while his mortal brother Castor was famous for his skill with horses. Both brothers voyaged in search of the Golden Fleece as Argonauts, and then fought in the Trojan War to bring their sister home to her husband Menelaus. They are traditionally depicted as armed with spears and riding a matched pair of snow-white horses.

The most common explanation for their presence in the heavens is that Pollux was overcome with sorrow when his mortal brother died, and begged Zeus to allow him to share his immortality. Zeus, acknowledging the heroism of both brothers, consented and reunited the pair in the heavens.

The stars of Gemini include two of the most recognisable in the heavens: the twins Castor and Pollux.

> *Castor (alpha Geminorum)* is the slightly dimmer star. It has a visual magnitude of 1.93 and is 52 light years distant. It isn't a particularly large star, at about twice the Sun's diameter. The star is a noted binary, discussed below.
>
> *Pollux* is the brighter of the two stars with a visual magnitude of 1.16 and a distance of 33.7 light years. It is also considerably larger, with an estimated diameter of about ten Suns.
>
> > Castor and Pollux are 4.5 degrees apart, which helps observers estimate separation distances

between other stars.

Epsilon Geminorum is a supergiant at about 30 Sun diameters. This star may be as far away as 950 light years, but the combination of visual and absolute magnitudes suggests a much closer star, at only 190 light years.

Zeta Geminorum is the most distant of the bright stars in this constellation, at over 1200 light years. This is a cepheid variable (see below).

Eta Geminorum is a red giant, about 50 times the size of the Sun, at a distance of 280 light years. It is a visual binary and a variable (details below).

Double stars:

Alpha Geminorum is a well-known binary with the companion currently (2000.0) at a PA of 65° and separation 3.9". The visual magnitudes are 1.9 and 3.0. There is some disagreement over the precise period of the companion; one observer has it at 420 years, another at 511. More recent measurements put the orbit at 467 years and the orbit we've prepared uses this revised value.

This was the first binary system that was so recognized, in 1802 (or 1803, accounts vary) by William Herschel. However there is considerable speculation that the star was a known double long before that, perhaps even a century before Herschel made his announcement.

The companion, Castor B, is also a spectroscopic

binary, with its companion revolving around Castor B every three days.

In fact, the entire system is comprised of six stars, including a red dwarf, Castor C, which slowly revolves around both Castor A and Castor B. This star is also a variable (and therefore cataloged as YY Gem).

Delta Geminorum: visual magnitudes 3.5, 8.2, PA 225°, separation 5.8". The period is estimated at 1200 years; the companion is an orange dwarf which may be difficult to resolve in smaller telescopes.

Eta Geminorum is a visual binary that takes some work to resolve; the companion is only 8.8 (primary is 3.3), the PA is 266° and separation 1.4". This is nearly a fixed binary, with very little movement.

Variable stars:

Zeta Geminorum is a cepheid variable, from 3.62 to 4.18 every 10.15 days.

Eta Geminorum is a semi-regular variable with an average period of 232.9 days. It ranges from 3.2 to 3.9.

R Geminorum is a Mira-type long-period variable, with large variation from 6.0 to 14.0 every 370 days. The 2000 maximum should arrive in mid October.

Deep Sky Objects:

The only Messier object in Gemini is *M35 (NGC 2168)*. This is an open cluster easily enjoyed in small scopes. It lies just 2.5 degrees northwest of eta

Geminorum.

This cluster is extremely attractive, with gently curving rows of glittering stars. Several hundred stars make up the group, which is perhaps 2500 light years away.

The Eskimo Nebula (NGC 2392) is one of the more distant nebulae at an estimated distance of 10,000 light years. There is a tenth-magnitude central star. If you do have a large enough scope, be prepared for anything: Burnham thought the Eskimo Nebula suggested "the classic and unforgettable features of W. C. Fields."

While you can locate this blue-green object in small scopes, it takes a very large telescope to see the "face" of this nebula, the eyes, nose, and mouth and the "fur collar" that gave it its name.

To find this rather small planetary nebula draw an imaginary line between kappa Geminorum and lambda Geminorum. Now draw a perpendicular line from delta Geminorum, and just about where this line meets the other one is where you'll find the Eskimo Nebula.

NOTES

Grus

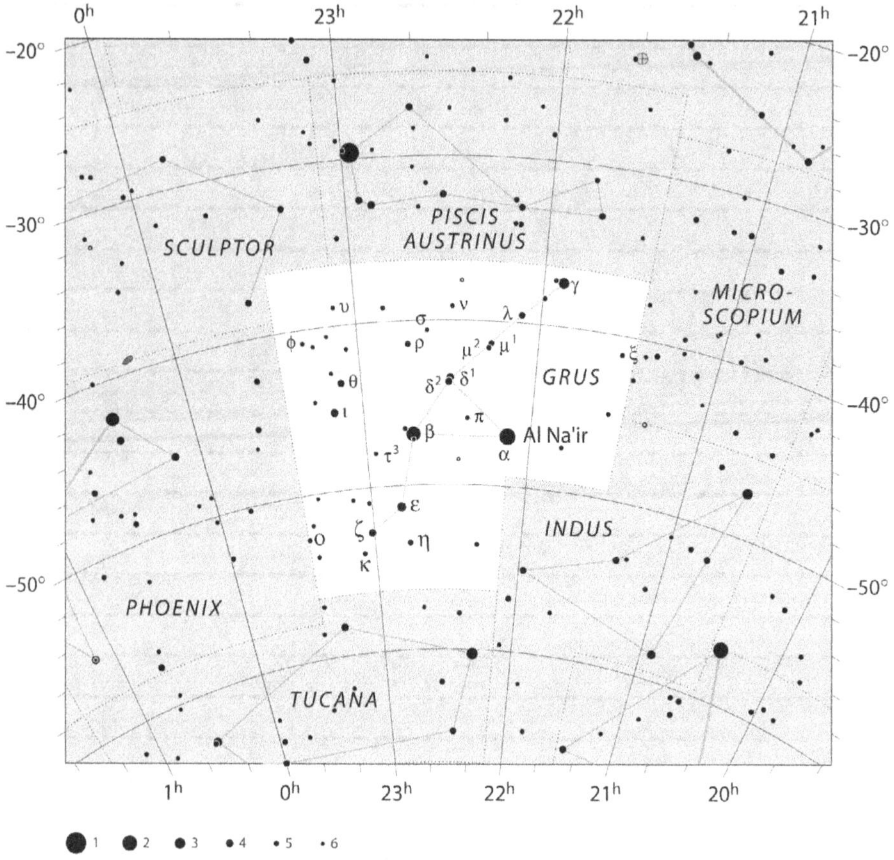

The Crane

𝕲rus lies just below Piscis Austrinus, and at one time was part of that constellation. Grus (The Crane) was so named by Johann Bayer, as listed in his 1603 star atlas.

The constellation is replete with very faint spiral galaxies, from eleventh to thirteenth magnitude, and a number of fine binaries, but little else.

The constellation is replete with very faint spiral galaxies, from eleventh to thirteenth magnitude, and a number of fine binaries, but little else.

Alpha Gruis is called *Al Nair*, or "The Bright One" (i.e., of the Fish's Tail), an obvious reference to its former association with Piscis Austrinus.

The star is almost as bright as Fomalhaut. The rest of the Bayer stars range from second to fifth magnitude.

Double stars:

Delta^1 Gruis and *delta^2 Gruis* form a pleasant optical binary of contrasting colours: yellow and red.

Theta Gruis: 4.5, 7; 75 degrees, 1.1".

Pi^1 Gruis: 6.5, 11; 201 degrees, 2.7".

Pi^2 Gruis: 6, 12; 214 degrees, 4.6".

Sigma^2 Gruis: 6, 10.5; 263 degrees, 2.7".

Tau^2 Gruis: 7.5, 8; 166 degrees, 0.3".

Upsilon Gruis: 5.5, 9; 211 degrees, 1.1".

Variable stars:

The variables of Grus are not very notable:

Beta Gruis is an Lc type irregular variable, with a range from 2.0 to 2.3.

Delta^2 Gruis is an Lb irregular variable, with a range from 3.99 to 4.2

Deep Sky Objects:

NGC 7213 is a spiral galaxy in the field of *alpha Gruis*, just 16' to the SE. This is one of the brightest of the many spiral galaxies in this constellation.

NOTES

Heracles

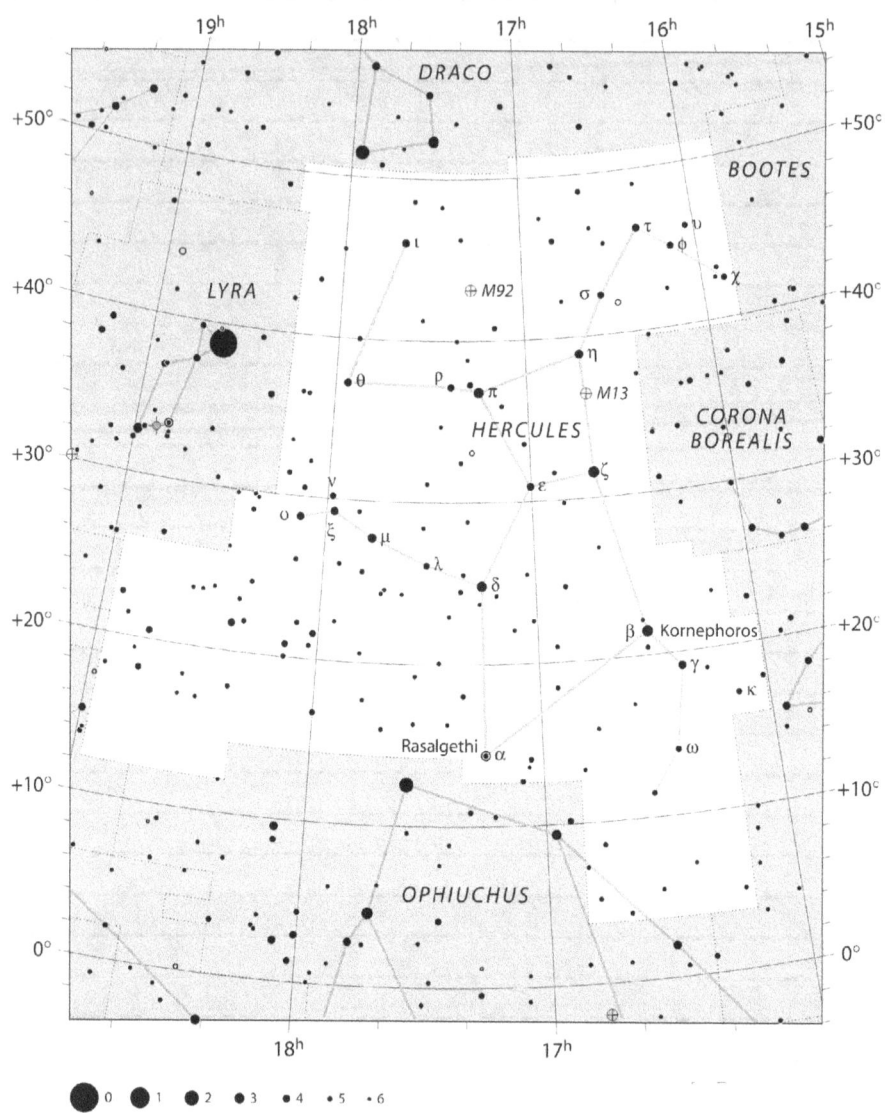

Heracles was perhaps the greatest hero in all mythology. He was the son of Zeus and Alcmena, and was hounded all his life by Hera. (This is deliciously ironic, because in the original Greek myths, Hera is named Hera and Heracles is Heracles, which means "glory of Hera.") Hera was unhappy with Zeus's infidelity, and saw Heracles as a living, breathing symbol of her shame. She delayed his birth, and when Heracles was a mere baby (but a big one!) sent two snakes into the crib he shared with his mortal half-twin Iphicles. Heracles killed them both with his bare hands, marking the beginning of his career as a monster-killer.

After a precocious childhood and adolescence, Heracles married Megara (daughter of Creon, king of Thebes). Hera succeeded in driving him mad, though, and he killed his wife and his children. As atonement, he serves the king Eurystheus, performing the twelve labors for which he is most famed.

Unless you are an avid stargazer, you might not be sure just where to look for Hercules. While the fifth largest constellation, it isn't very obvious.

And yet Hercules boasts one of the finest collection of binary stars, and two Messier objects as well.

We will make a fine distinction here: the constellation name is *Hercules,* while the Greek hero is *Heracles*.

Heracles was named after the greatest of Greek goddesses, Hera. Her name means "Lady" and she was the daughter of Cronus, and sister of Zeus (they were twins). Zeus later changed into a cuckoo and seduced his sister (he had that kind of reputation), and the two were married.

Hera became the Queen of the Heavens: goddess of childbirth, marriage, and of women, she was the most widely beloved of goddesses in antiquity. It would only be natural that the greatest of Greek heroes would be named after her: Heracles means "the glory (or honour) of Hera".

Although named after Hera, Heracles didn't have her immediate respect. Heracles was the son of Zeus and a mortal woman (Alcmene). Hera resented Zeus' philandering nature, and tried to have the child killed. She sent two monstrous snakes to his crib, but the infant strangled them both with his pudgy little hands.

Heracles became a favorite with the gods. Apollo made his bow and arrows; Athene gave him a magnificent robe; Hermes provided him with a sword, and Castor (the greatest warrior) taught him how to use it. Hephaestus, the smithy of the gods, made a golden breastplate for Heracles. Thus armed and protected, Heracles paraded through Greek mythology, performing eight heroic deeds and the Twelve Labors.

In fact, the very word "hero" has links with the names Hera and Heracles. The Romans would change his name to Hercules (and hers to Juno, and Zeus to Jupiter).

"Hercules" came to Italy in his tenth labour. He would later be given credit for abolishing human sacrifice in the land.

The constellation was originally represented as a kneeling man, with a foot on the neighbouring dragon (Draco). Some star names reflect this earlier association.

Hercules is a sprawling constellation just to the west of Lyra. From Vega (alpha Lyrae) swing to the west-southwest eight degrees. This is *theta Herculis*, a 3.86 magnitude star - which is about typical brightness for the main stars of this constellation.

The principal stars are found farther south. Star hop from theta over to pi Herculis, and then to the southwest (about the same distance from pi Herculis to Vega) is *beta Herculis*, which is actually the brightest star in the constellation.

Now look southeast and you will come across alpha Ophiuchi (Ras Alhague), at 2.1 magnitude, the brightest star of the region. Alpha Herculis is northwest of this star.

> *Alpha Herculis* is better known as *Ras Algethi: The kneeler's head.* It is estimated to be from 430 to about 650 light years. Some authorities believe the star to be as large as 400 solar diameters.
>
> This is a fine double: a red (or orange) supergiant and a blue-green giant (see below). The primary is also an irregular variable (see below).

Double stars:

Hercules has several binaries with contrasting colours, as well as several close binaries, challenging those with larger telescopes.

> *Alpha Herculis* is a visual binary with a very long period, something like 3600 years. 3.2, 5.4; PA 104, separation 4.6".
>
> *Zeta Herculis* is a rapid binary with colour contrast, a

yellow primary and red companion with a period of 34.45 years: 2.9, 5.5. The 2000 values: PA 12° degrees, and the separation 0.7".

Kappa Herculis is an easily resolved binary: 5.3, 6.5; PA 12 degrees, separation 28.4".

Rho Herculis: two white stars which make a lovely double. 4.6, 5.6; PA 326, separation 4.1".

95 Herculis is a very attractive double with contrasting colours, often described as gold and silver (although you may disagree): 5.0, 5.1; PA 258 degrees, separation 6.3".

99 Herculis is a very close rapid binary: 5.1, 8.4; currently the PA is 92 degrees and the separation 0.3".

100 Herculis is another gorgeous binary of two equal white stars easily resolved. 5.9; 5.9; PA 183 degrees, separation 14.2"

Struve 2319. This is a very beautiful binary of two rather faint stars: 7.2, 7.6; PA 191 degrees, separation 5.4".

Variable stars:

Alpha Herculis is an irregular variable with a range from 2.7 to 4.0, with a period of roughly three months.

S Herculis is the brightest long-period variable in Hercules, with a visual magnitude range of 6.4-13.8 every 318.14 days. The maximum for the year 2000 should occur in mid July.

Deep Sky Objects:

There are two Messier objects in Hercules: M13 and M92.

M13 (NGC 6205) is a spectacular globular cluster sometimes known as "The Hercules Cluster". It is universally acclaimed as the best globular in the northern hemisphere.

> This is a very compact cluster of over a million stars. It is also very old - at an estimated age of ten billion years. It's around 25,000-30,000 light years away.
>
> M13 lies on a line between eta Herculis and zeta Herculis, due west of pi Herculis.

M92 (NGC 6341) is also a globular cluster, located nine degrees northeast of M13, and six degrees directly north of pi Herculis.

> M92 is also very striking and worthy of consideration, even if considerably overshadowed by M13.

NOTES

Horologium

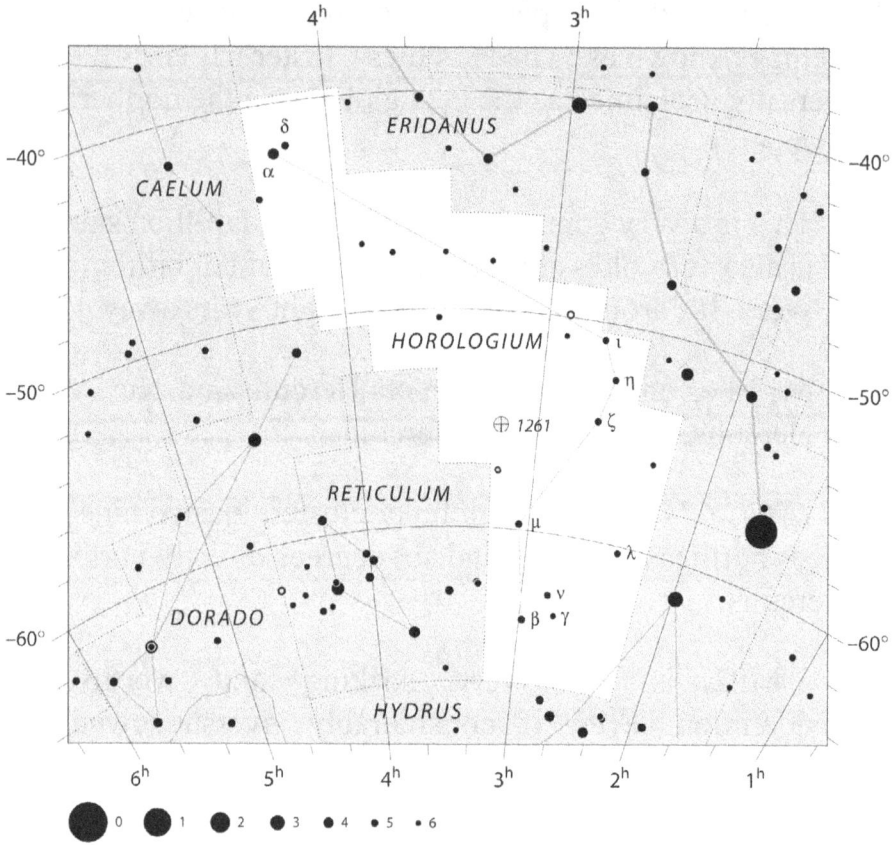

The Clock

It was created in the 18th century by Abbé Nicolas Louis de Lacaille, who originally named it **Horologium Oscillitorium** after the pendulum clock to honor its inventor, Christiaan Huygens. The name has since been shortened to be less cumbersome.

Like the majority of Lacaille's constellations, Horologium is a curious asterism, and has very little to offer. There are no binaries of any interest here, and no deep sky objects. One Mira-type variable (R Hor) is a rather interesting star.

Its Bayer stars are few and rather faint, ranging from 3.9 to 5.4

Variable stars:

R Horologium is a Mira-type variable with a very wide range. Indeed, this star has one of the largest ranges known: from a fairly bright 4.7 to an extremely faint 14.3 visual magnitude. The period is 407.6 days. In the year 2000 the maximum should be achieved near the end of September.

The star is two degrees NE of iota Horologii

NOTES

_____
_____
_____
_____
_____
_____

Hydra

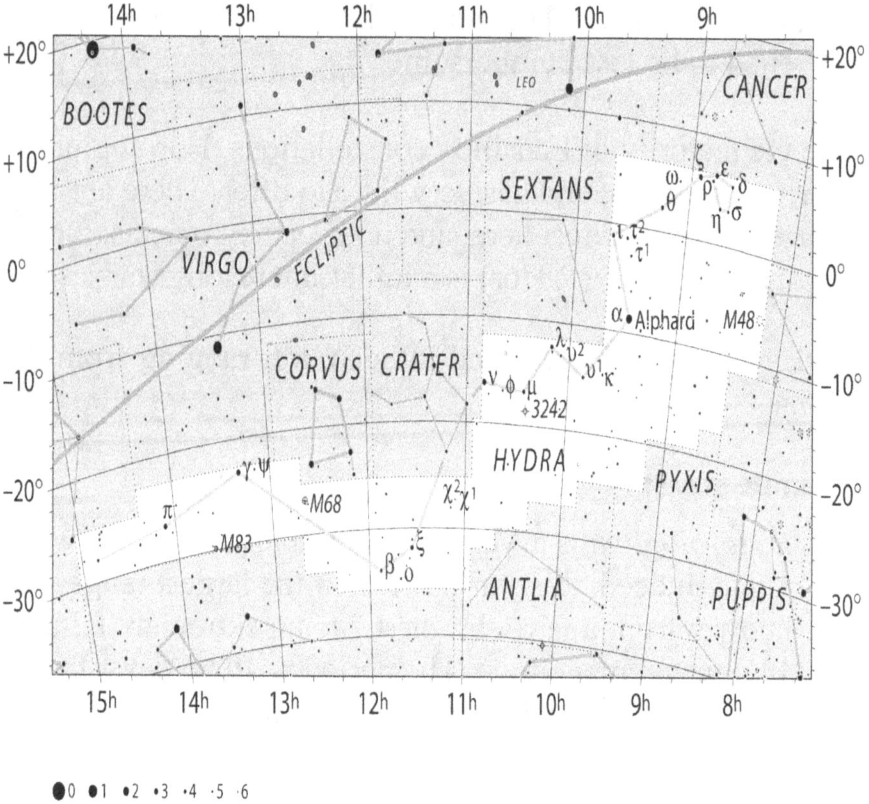

Water-Snake

This constellation represents the Lernaean Hydra, slain by Heracles as his second labor. The Hydra was a multi-headed monster--according to Diodorus (first century B.C.), it had a hundred heads; Simonides (sixth century B.C.) said it had fifty. The most common opinion, however, seems to be that it had nine. What made the Hydra so difficult was the fact that, whenever one of its heads was chopped off, two would grow in its place. Heracles managed to get around this rather major obstacle by having his nephew, Iolaus, cauterize each stump with a hot iron as soon as Heracles could chop off a head. The hero then buried the monster's immortal head beneath a rock. The task was made somewhat more difficult by Hera, who sent a crab to nip at the feet of Heracles while he battled the Hydra.

Hydra is a long and wandering constellation, stretching almost from Canis Minor to Libra. It lies south of Cancer, Leo, and Virgo, and is best seen in the northern hemisphere during the months of February through May.

As a constellation Hydra sinuously winds down from the northern hemisphere, bordering Cancer, to as far in the southern hemisphere as Centaurus, stretching out to around one hundred degrees in the process. At one point its body is actually cut off by another constellation (the southwestern tip of Crater).
Although the constellation is best seen in March and April, its brightest star, *alpha Hydrae*, transits on 12 February. Slowly over the next few months the Hydra then slithers over the skies earlier and earlier.

To find Hydra, locate Regulus *(alpha Leonis)*. Now drop straight down twenty degrees, passing through the uninspiring constellation of Sextans. The bright star to the west is *Alphard (alpha Hydrae)*.

You've just found the "heart" of the Hydra. To find its head, look for the compact stars to the northwest. It is easiest to do this with the naked eye, or with binoculars. These stars are not quite as far north as Regulus.

The tail of the monster stretches far to the east, even past Spica (alpha Virginis). You'll have to wait until the small hours of the morning to see these stars, or until later in the spring when they appear at a more reasonable hour.

There are a number of interesting deep sky objects, including two Messier objects, and several very nice double stars. All in all, Hydra's stars are of average brilliance, except for *Alphard (alpha Hydrae)*.

> Alphard is a bright giant, about 25 times the Sun's diameter. Its distance is 89 light years, and it has a luminosity of 95 Suns.
>
>> The Arabs called the star "Al Fard al Shuja" (The Solitary One in the Serpent) while later on Tycho Brahe named it "Cor Hydrae" (The Hydra's Heart).
>
> *Beta Hydrae* is the southern-most star of any importance. Quite dim for its name, the star is a visual binary (see below).

Double stars:

> *Beta Hydrae* is a pair of nearly equal stars (4.5, 5) at PA 8° and a separation of 0.9".

Epsilon Hydrae is a multiple binary; four stars can be seen and another has been calculated to exist.

> Components A and B form a rapid binary with a period of 15.05 years; its orbit is nearly circular. Presently (late February 1996) the companion star has a PA of 166° and separation 0.26".

> Component C is much easier to resolve, with a period of 990 years. At present it can be found at a PA of 298.5° and separation 2.7".

Chi^1 Hydrae is a binary of two similar stars (5.8, 5.8) with an even more rapid orbit. Its period of 7.4 years means an exceedingly difficult binary to resolve. If you've a large enough telescope, you'll find the companion at these values in late February 1996: PA 31° and separation 0.046".

Sigma 1474 is a fixed binary forming a wonderful triple. AB: 6.8, 7.9; 24° and separation 70"; C: 6.9, 23°, 76" separation.

> To find the binary, locate *nu Hydrae* then move one degree northwest. (Just north half a degree is the nearly attractive Sigma 1473.)

Variable stars:

R Hydrae ranges from a visual magnitude of a fairly bright 3.5 to a faint 10.9 every 388.87 days. However this star has shown a decrease in its period. (In 1920 it had a 404 day period.) In 2000, based on its current period of 388 days, the maximum should be attained during the first week of August.

This star is one of the earliest variables to be catalogued, having been discovered in 1704. Only *Mira (omicron Ceti), beta Persei,* and *chi Cygni* predate R Hydrae.

Burnham gives a finder's chart. Note that the brightest star in the region, SS Hydrae, is also a variable, so don't base R Hydrae's visual magnitude on this star.

T Hydrae has a shorter period, 298.7 days, and a range of 6.7-13.5. To find T Hydrae, star hop westward from Alphard. First, due west of alpha Hydrae you find 24 Hydrae, then 20 and 19 (which is the brighter of the two). Now just about the same distance that you've taken to get to 19 Hydrae you'll find T Hydrae, due south of 17 Hydrae one and a half degrees.

This star, 17 Hydrae, is the brightest star in the region, and a visual binary as well (7, 7; 359°, 4.3"). The star is useful to determine the visual magnitude of T Hydrae to the south.

In 2000 T Hydrae should attain its maximum brightness in mid October.

Deep Sky Objects:

Hydra has three Messier objects: *M48, M68,* and *M83.*

M48 (NGC 2548). Messier actually gave the wrong location for this star cluster, putting it four degrees north of the current position. But by his description this seems to be the right object.

Not terribly spectacular, this is a group of fifty stars, the brightest of which is about 8.8 visual magnitude. The cluster is thought to be about 1700 light years away, and is easily seen in binoculars or small telescope.

M68 (NGC 4590) is a much richer globular cluster of over a hundred thousand stars, resolved in medium-sized telescopes.

> The cluster lies in a desolate part of the sky. Locate gamma Hydrae then move west to beta Corvi. Now drop down three degrees to the brightest star in this region, a fifth-magnitude star (this is the binary B230: 5.5, 12; 170 degrees, separation 1.3"). M68 is about a half degree to the northeast.

M83 (NGC 5236) is a spiral galaxy sitting on the Hydra-Centaurus border farther to the east, nearly twenty degrees south of Spica (alpha Virginis).

> While Burnham says this is considered one of the brightest galaxies with a visual magnitude of about 8, other references give it only a 10. And since it is very diffuse, it may be difficult for those living in northern latitudes.

NGC 5694 is an extremely compact globular star cluster, thought to be in the region of 100,000 light years away.

> The cluster sits just east of the mid-way point between *pi Hydrae* and *sigma Librae*, at the

border with Hydra. From *pi Hydrae* move east until you encounter a group of five magnitude stars lined up roughly north-south. These are 54, 55, 56, and 57 Hydrae. NGC 5694 lies one degree west of 57 Hydrae.

NGC 3242 clearly deserves to be a Messier object. Small but bright, at a visual magnitude of 8.6, this planetary nebula is often called *The Ghost of Jupiter* because of its slight resemblance (?) to that planet. Also at times called *The Eye Nebula*, perhaps a closer description.

> The nebula gives off a soft bluish colour, sometimes described as a "glow", that is clearly visible even in small scopes. The central star may be difficult to resolve: this is an 11.4m star, a blue dwarf considered to be as hot as 60,000 kelvin.

> The nebula is one of the easiest to find. Just locate *mu Hydrae* then move south two degrees.

> Trying to resolve the inner ring could prove difficult. Large telescopes should show the object as resembling an eye, with the central star the pupil. The greenish-blue colour adds to this intriguing sight.

NOTES

Hydrus

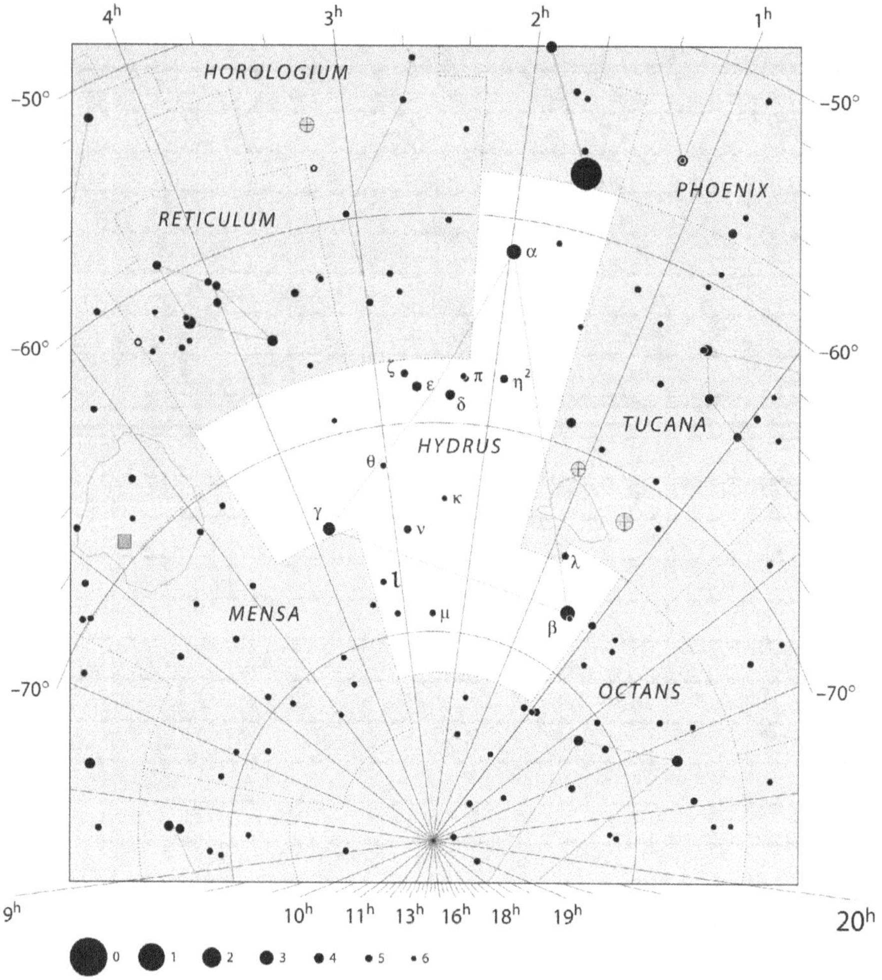

The Lesser Snake

Hydrus, "The Lesser Snake", was one of Johann Bayer's constellations, found first in his 1603 publication Uranometria. It was meant to be the southern hemisphere's answer to Hydra, but it has far fewer objects of interest.

Hydrus is a rather stiff snake, perhaps mostly resembling a cobra, with its head erect and body curled. As with most of these obscure constellations, its Bayer stars are far from complete, and fairly faint.

Binary stars:

h3475 is a fine binary about one and a half degrees just north (and slightly west) of alpha Hydri.

NOTES

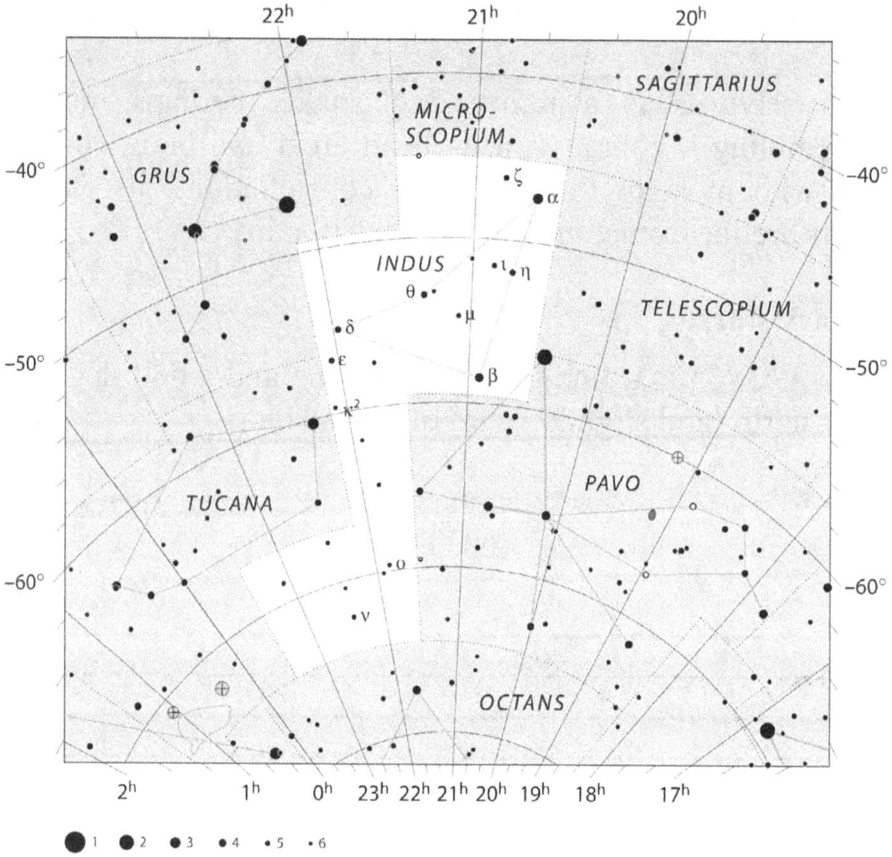

American Indian

Johann Bayer wanted to honour the American Indian in his collection of new constellations for his 1603 book *Uranometria*. Indus is the result, a collection of stars ranging from 3.1 to 5.3 visual magnitude.

The constellation has little more than a few binaries, and one of the closest stars to our Solar System.

As it closely matches our own sun, *Epsilon Indi* has been studied as a possible candidate for planets, however none have yet been found circling the star.

Binary stars:

Alpha Indi is a wide binary with a very faint companion: 3.1, 12.5; PA 199 degrees, and separation 67.4".

Delta Indi is an extremely rapid binary, only 12 years; thus the component is very close. Epoch 2000 values: 4.4, 5.5; PA 50 degrees, separation 0.2".

Theta Indi is a pleasant binary, easily resolved: 4.4, 7.0; PA 275 degrees, separation 6".

NOTES

Lacerta

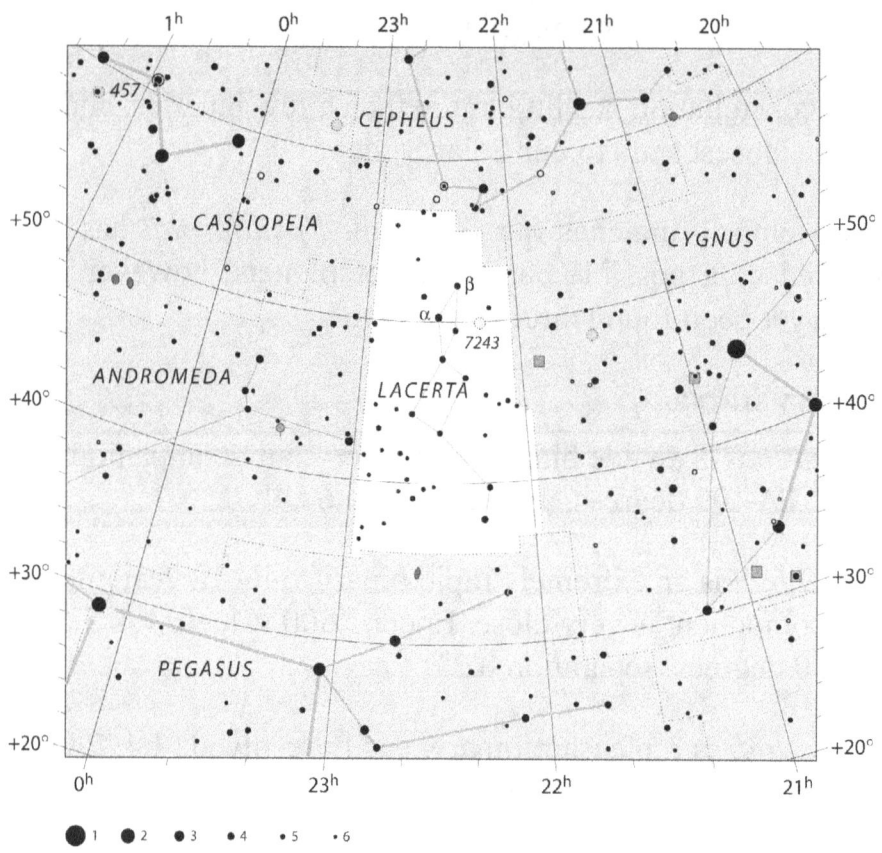

The Lizard

Lacerta, is one of seven constellations introduced by Johannes Hevelius.

Born in Gdansk, Poland, 28 January 1611, Hevelius died the same day in the same place 76 years later.

Hevelius is mostly known for his atlas of the Moon (*Selenographia*, 1647). His star catalogue of 1564 stars was the most complete up to that time. It was in this catalogue and an accompanying celestial atlas (*Prodromus Astronomiae*), both published posthumously in 1690, that Hevelius introduced seven new constellations.

Hevelius had built his own observatory, on the roof of his house, as well as a number of quality telescopes. His stellar observations were the most accurate to his time, and for that reason the celestial atlas was a remarkable achievement.

It was John Flamsteed who -- in his own star catalogue published posthumously in 1725 -- popularised Hevelius' new constellations: Canes Venatici, Lacerta, Leo Minor, Lynx, Scutum, Sextans, and Vulpecula.

Due to the fact that the stars in Hevelius's constellations were borrowed from neighbouring constellations, many of the Bayer (Greek labeled) stars are missing. Thus Lacerta, "The Lizard", has only two Bayer stars.

Lacerta lies between Cygnus and Andromeda. It has a few fine binaries and several nice deep sky objects.

Binary stars:

8 Lacertae is a multiple system with quite wide components; these are the three brightest components:

> AB: 5.7, 6.5; PA 186 degrees, separation 22.4".
> AD: 9.3; PA 144 degrees, 81.8".
> AE: 7.8; PA 239 degrees, 336.6".

Struve 2902 is the most attractive binary in Lacerta: 7.6, 8.5; PA 89 degrees, separation 6.4".

> The binary is 1.5 degrees SE of 2 Lacerta, along a line between 2 Lac and 6 Lac.

h1823 is a fine multiple system 1.5° northeast of 12 Lac.

> AB: 6.8, 12.5; PA 259 degrees, separation 19.2"
> AC: 8.5; PA 338 degrees, 82.1".
> AE: 8.9; PA 263 degrees, 18.3".

Variable stars:

8^A *Lac* is a BE variable (shell star) with unknown range and a period of about 16 to 18 years.

> *12 Lac (DD Lac)* is a beta Cep type variable: 5.16 to 5.28 every 4h38m3.2s.

Deep Sky Objects:

NGC 7209 is an open cluster of fifty stars ranging in visual magnitude from nine to twelve. The cluster is 2.5° west of 2 Lac.

> NGC 7243 is another open cluster of forty or so

stars; the brightest star here is a fine binary (*Struve 2890*: 8.5, 8.5; PA 11, 9.4"). The cluster is 2.5° WSW of alpha Lac.

NOTES

Leo Major

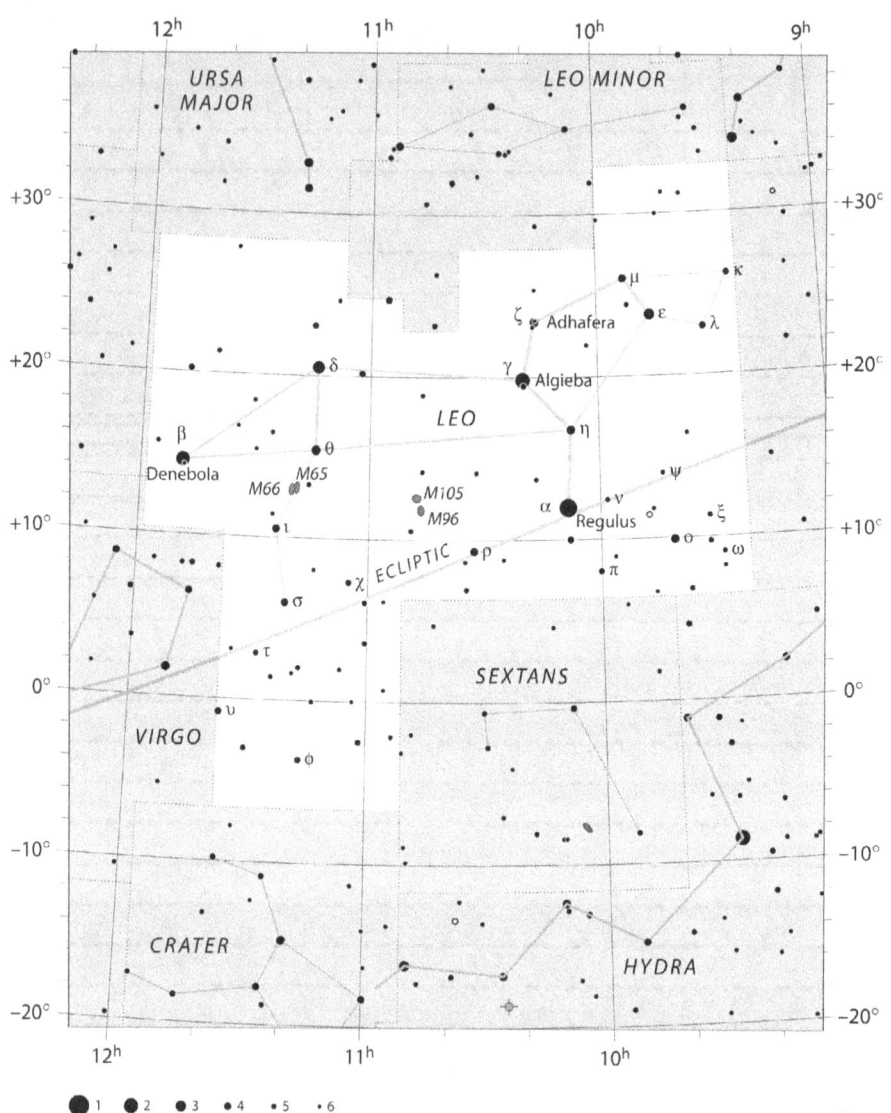

The Lion

The first on the list of Heracles' labors was the task of killing the Nemean Lion, a giant beast that roamed the hills and the streets of the Peloponnesian villages, devouring whomever it met.

The animal's skin was impervious to iron, bronze, and stone. Heracles' arrows harmlessly bounced off the lion; his sword bent in two; his wooden club smashed to pieces. So Heracles wrestled with the beast, finally choking it to death. He then wrapped the lion's pelt about him; it would protect him from the next labour: killing the poisonous Hydra.

As the story goes, the lion found its way to the heavens to commemorate the great battle with Heracles. Yet this isn't all there is to the story. For even in antiquity, long before the Greeks began telling stories, the lion was an ancient symbol of power.

Approximately three thousand years before the Christian era carvings and sculptures showed kings flanked with rampant lions. Indeed, the archaeological evidence suggests that at about this time the lion had already replaced an even earlier "sacred" symbol, the bull.

It has been suggested that this transfer of power from a horned animal to the lion was a change-over from a lunar-based to a solar-based religion. That is, instead of drawing their inspiration from a night-time symbol with a monthly cycle--a symbol which dealt with the fecundity of the earth and of its animals--the new rulers identified with an animal of strength and power, and with a heavenly body that ruled the day. Thus, as the bull had been identified with the moon, the lion was now associated with the sun. To assert this new religion, or new political structure, the lion was made to kill the bull. Its place in the heavens was therefore critical.

An intriguing theory, put forth thirty years ago by Professor Willy Hartner, eloquently describes the result. Briefly put, at about 4000 BC, the Lion is seen chasing the Bull over the horizon, announcing the end of winter and the beginning of spring. I shall quote Professor Hartner's descriptive words:

> *"The constellation Leo would have been directly overhead, standing at zenith and displaying thereby its maximum power [as it] kills and destroys the Bull trying to escape below the horizon, which during the subsequent days disappears in the Sun's rays to remain invisible for a period of forty days, after which it is reborn, rising again for the first time (March 21) to announce Spring equinox."* [See W. Hartner, "The Earliest History of the Constellations in the Near East and the Motif of the Lion-Bull Combat" *JNES* 24(1965)1-16.]

Thus Leo, slayer of Taurus, dominated the summer skies, the time that the sun passed through this constellation. Due to precession, the sun currently passes through Leo at the end of summer, from mid-August through mid-September.

Leo is a fairly compact constellation and, unlike so many other constellations, it is readily recognizable. *Alpha Leonis* is named "Regulus" because it was seen as the Heaven's Guardian, one who regulated all things in the heavens. While the name Regulus was given us by Copernicus, the star was better known in antiquity as *Cor Leonis*, the Lion's Heart.

> Regulus is a multiple binary, discussed below. Also, because Regulus lies so close to the ecliptic, the moon often passes close by, and even occults the star on very rare occasions.

Like other ancient constellations, many of the stars in Leo are named.

> *Beta Leonis* is called "Denebola": the Lion's Tail.

> *Gamma Leonis* is "Algeiba", Arabic for forehead, but more correctly named *Juba*, meaning mane.

> *Zeta Leonis* is "Aldhafera", the meaning is uncertain;

> *Epsilon Leonis* and *mu Leonis* go under the name of "Al Ashfar", the eyebrows.

> *Delta Leonis* is "Zosma", a Greek word meaning girdle.

Lambda Leonis is Alterf, apparently meaning "extremity". It's located right at the tip of the lion's mouth.

Double stars:

> *Alpha Leonis (Regulus)* is a multiple system. Component B is very wide: (8.1m, PA 307 degrees, 177"), and this star has its own companion ("C"), a very faint 13m dwarf, with a period of about 2000 years, now approximately 2.6" and a PA of about 86 degrees.

> A fourth companion, D, is only optical. That is,

there is no gravitational bond with the others, but before that was established, it too became a part of the group. It is found at 274 degrees, and 217".

Gamma Leonis is a notable binary with a slow orbit. While Burnham lists three possible periods (407y, 701.4, and 618.6) we have settled on the latter as the most probable, and based its orbit on this period.

> Presently the companion is very gradually drawing away from the primary. The current values are: PA 124 degrees and separation 4.4".

Iota Leonis is a more rapid binary, with a period of 192 years. Its orbit shows that the 6.7m companion is slowly increasing its distance (now at PA 122 degrees and separation 1.62").

Variable stars:

> *R Leonis* is the only variable of note in Leo. This isn't your typical Mira-type long-period variable. First of all, it's usually a very faint 11.3m star, which grows to an extremely bright 4.4m every 309.95 days. In 2000 the maximum should arrive in the last week of February.

Secondly, its color is an unusually deep red, approaching purple. Surrounded by a number of white stars (18, 19, 21 Leo.) its own color is even more pronounced. Thus *R Leonis* has become a favorite subject for many variable star observers.

Deep Sky Objects:

Leo has five Messier objects: *M65*, *M66*, *M95*, *M96*, and *M105*.

M65 (NGC 3623) and *M66* (NGC 3627) make a splendid pair of spiral galaxies in the same field, between theta Leonis and iota Leonis.

> This is a fine binocular duo, or use a small telescope. M66 is the one to the east. Both galaxies are elongated north-south; M65 has a tighter spiral and is perhaps the more noticeable.
>
> About a degree north, hovering just between M65 and M66, is NGC 3628, a galaxy seen edge-on. Actually this is larger than either Messier object, but much dimmer because it is seen edge-on.

M95 (NGC 3351) and *M96* (NGC 3368) form another nice pair, although farther apart. The two are found is a group of galaxies midway between alpha Leonis and theta Leonis, and just slightly to the south.

> Of the two, M95 is to the west. This is a curious round object, with a very faint circular bar. M96 is a tight spiral galaxy, much brighter than its neighbour. Both this pair and M65/M66 are considered to be about 30 million light years away.

M105 (NGC 3379) is a much dimmer galaxy to the north-north-east of M96. Along with NGC 3384 and NGC 3389, which lie just to the east, this object forms a small triangle of galaxies.

Then there is *NGC 2903*, which somehow escaped Messier's telescope. This deep sky object is judged to be a visual magnitude of 8.9, which makes it brighter than any of the above Messier objects, and covering a larger area as well. It is an elongated multiple-armed spiral located directly south of lambda Leonis, one and a half degrees.

Indeed, there are many more galaxies in Leo to explore. Most of them lie between alpha and beta Leonis, with a smaller group scattered around gamma Leonis. Most of them are 10-12m, so the larger the telescope the more favorable the viewing.

If you wish a real deep sky challenge, try *Wolf 359*. This is an extremely faint red dwarf, and the third closest star, at 7.65 light years. It has a visual magnitude of only 13.53, which renders it all but lost among the millions of other stars. Only as large as Jupiter, it has a luminosity about 1/65,000 of the Sun's; its absolute magnitude is calculated at 16.7m.

Its Epoch 2000 values are: right ascension 10h 56m, declination 07 degrees, one second. If using Tirion's *SkyAtlas 2000.0*, while this chart doesn't show the star, you can easily find the region. Locate 56 Leo (west of sigma Leonis) then place a mark on the ecliptic just above this star. (The ecliptic is the dotted line running north of this star). This is where you'll find *Wolf 359*. Now you'll need Burnham's finder (on his page 1072), a nice dark sky, and plenty of patience.

The Leonid Meteor Showers

The Leonids are meteor showers which appear every 16-17 November, so called because their radiant point is in the

Sickle of the Lion.

While Leo is only visible very early in the morning at this time of year, the meteors stream across the west, through Hydra, Canis Minor, and even Orion.

The meteors are debris from the Tempel-Tuttle comet, which orbits the sun every 33 years. While the showers usually bring 10-20 meteors per hour, three times a century the meteor shower is particularly dense, several hundred per hour with a quick burst of 1000 or more per hour.

These denser showers usually occur in years ending in 33, 66, and 99. The 1966 display was one of the better, with around 5000 meteors in one span of twenty minutes. By contrast, 1999 saw a very brief burst of around 2000 meteors as recorded by observers in the Middle East, while North American residents would only view several dozen per hour.

NOTES

Leo Minor

Lesser Lion

Leo Minor is one of the seven constellations introduced by Johannes Hevelius in his posthumous catalog of 1690. The others he introduced are Canes Venatici, Lacerta, Lynx, Scutum, Sextans, and Vulpecula.

While Lacerta has two Bayer stars (alpha and beta), Leo Minor has only one; curiously it's Beta LMi. It isn't even the brightest star of the constellation; *46 LMi* has that honour with a magnitude of 3.8.

Leo Minor is the only northern hemisphere constellation which has no "alpha". The others, all in the southern hemisphere, are Norma, Puppis, and Vela.

Leo Minor lies just above Leo and to the right of Ursa Major, with a nondescript asterism; it's a "lesser lion" in name only.

The constellation has one binary of any interest, one Mira variable, and a few galaxies, including a rare interacting pair of galaxies.

Binary stars:

Beta Leo Minoris is an extremely close binary with an orbit of 39 years: PA 223°, separation 0.4".

Variable stars:

R Leo Minoris is a Mira type variable ranging from 6.3 to 13 every 372.2 days. In 2000 the maximum should occur in early September.

Deep Sky Objects:

NGC 3003 is an almost-on-edge galaxy, quite large but

rather faint. It is four degrees SW of 21 LMi.

NGC 3395 is one of two interacting galaxies, two small and bright galaxies which seem to exist in some kind of symbiotic relationship. The two are found 1.5 degrees SW of 46 LMi.

NGC 3396 is the other half of the interacting galaxies, with a separation of 1.7'. It is slightly fainter (12.8) and slightly smaller.

NOTES

Lepus

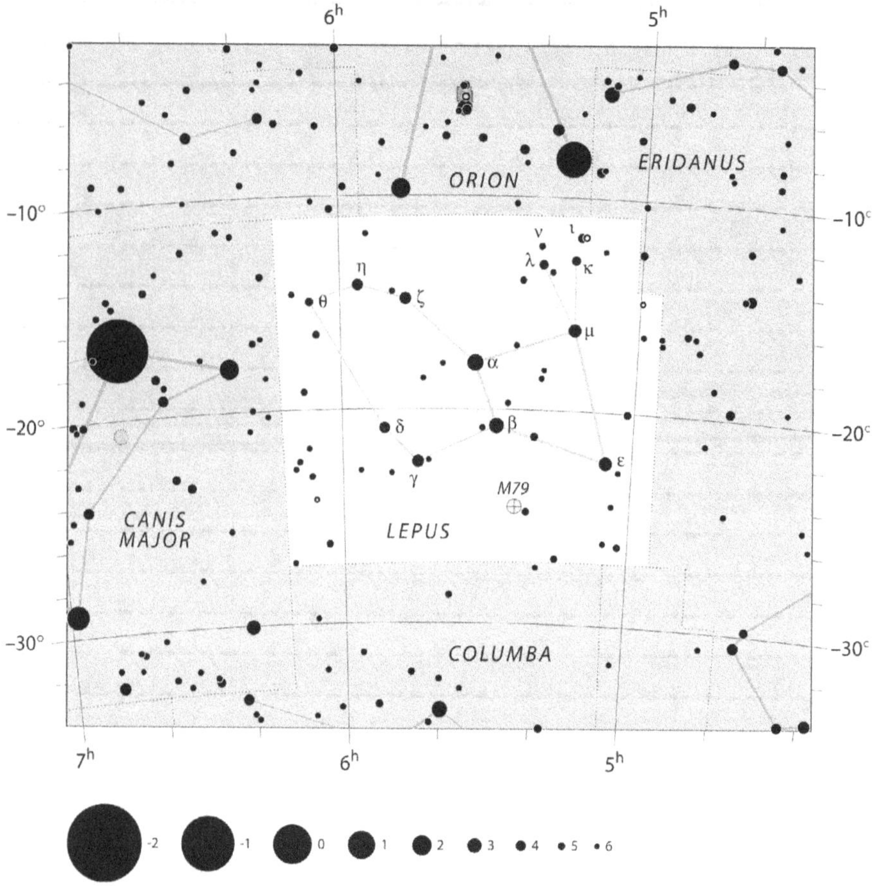

The Hare

Lepus, "The Hare", is an ancient constellation found under the feet of Orion, the Hunter. No one seems to know just which culture first saw the constellation as an animal; the Arabs saw it as the "throne of the central one" (i.e. Orion).

Lepus, The Hare is not to be confused with Lupus, The Wolf, which is a spring constellation.

Lepus is often ignored, as Orion is such a dominating constellation. Yet Lepus contains a number of interesting objects. Its Bayer stars are generally third and fourth magnitude.

Double stars:

Beta Leporis is a close binary with faint companion: 2.8, 11; PA 330 degrees, separation 2.5".

Gamma Leporis is a wide binary with slight colour contrast, yellow and orange (although observers vary): 3.7, 6.3; PA 350, separation 96.3".

Kappa Leporis (Struve 661) is a fixed system: 4.5, 7.4; PA 358 degrees, separation 2.6".

h3750 is a fixed binary: 4.7, 8.5; PA 282 degrees, separation 4.2".

h3752 is a fine multiple in the same field as M79.

AB: 5.5, 6.5; PA 97 degrees, separation 3.1". AC: 9; PA 106 degrees, separation 59".

h3780 is a noted multiple system which also goes under the name NGC 2017.

AB: 6.4, 7.9; PA 146 degrees, separation 0.8".
AC: 8.5; PA 136 degrees, separation 89.2".
AE: 8.4; PA 7, separation 76.1".
AF: 8.1; PA 299, separation 28.8".
AG: 9.5; PA 49, separation 59.8"

Variable stars:

Mu Leporis is an alpha CV type variable: 2.97 to 3.41 about every two days.

Rho Leporis is an alpha Cygni type variable: 3.83 - 3.90.

R Leporis is a long-period (Mira) variable that ranges from about 6 to about 11.5 every 427.07 days. However sources vary over this figure, and you will find quoted a period ranging from 427 to 440 days. In 2000 the maximum may occur in the last week of December, depending of course on the star's current period.

> The star glows with an unusually intense red; it goes by the name of *Hind's Crimson Star* since John Russell Hind (in 1845) wrote that it resembled "a blood-drop on the background of the sky". Although, unfortunately, as the star brightens it loses much some of its intense colour.
>
> The star is 3.5° NW of mu Leporis. Burnham (p.1094) has a finder's chart. (The star less than two degrees south of R Lep is the close binary b314.)

Deep Sky Objects:

Lepus has one Messier and a tiny star cluster which is

actually the half dozen stars which go to make up the multiple h3780.

M79 (NGC 1904) is a small globular cluster about 3.5 degrees SSW of beta Leporis. In the same field, half a degree to the SSW of this cluster, is h3752 (see above).

NGC 2017 is a group of a half dozen stars, all gravitationally bound (h3780, see above). The "cluster" is found seven arc minutes due east of alpha Leporis.

NOTES

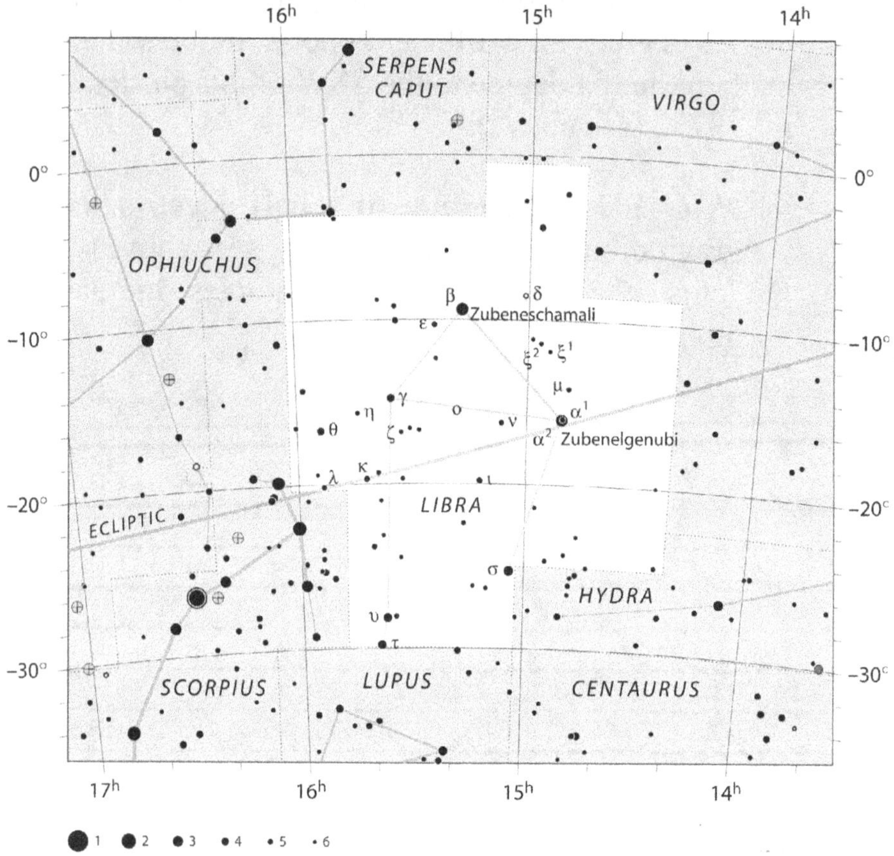

The Scales

Libra means "The Scales" or "Balance", so named because when the zodiac was still in its infancy, some four thousand years ago, the sun passed through this constellation at the autumnal equinox (21 September). At the two equinoxes (Spring and Autumn) the hours of daylight and darkness are equal.

As a symbol for equality, the constellation came to represent Justice in several middle Eastern cultures. However, the Greeks had a different perspective; at one time Scorpius, which lies just to the east, was much larger, and the stars that make up Libra were then known as the Claws of the Scorpion.

Eventually, however, these stars of Libra came to represent the Golden Chariot of Pluto. The story of Pluto's abduction of Persephone is a widely known Greek myth, perhaps because it has such a strong astronomical association.

Pluto's (or Hades') Golden Chariot was used whenever the Lord of the Underworld wished to visit the Upperworld, usually to seduce a nymph. But when he took Persephone back to Tartarus, the deepest part of Hades, the Upperworld would change forever.

> The name of the ruler of the Underworld was actually Hades. Hades was a brother of Zeus and of Poseidon; he was usually ignorant of the happenings of the Upperworld, only emerging rarely from his dark kingdom.
>
> Deep beneath the earth, he owned all its mineral riches, but his favorite possession was a gift from the Cyclopes: a helmet that rendered him invisible. (Those familiar with Wagner's *Ring Cycle* will recognize the leitmotif, and a number of others in this story of

Persephone.)

It was considered imprudent and dangerous to mention the names of certain gods and goddesses. Thus the Furies, or Cronies, were called Eumenides (Kindly Ones), and Hades was called Pluto (Rich One).

The Bayer stars are fairly dim, except for two two-magnitude stars, *alpha^2* and *beta*. The constellation has several objects of interest, including some fine double stars and an unusual variable.

Alpha Librae is also known as *Zubenelgenubi*, a derivation of an older Arabic name that translates into "Southern Claw" (i.e. of the Scorpion). The star is a wide binary of unequal stars (see below).

Beta Librae is called *Zubeneschamali*, "The Northern Claw". This white star has been described by some to be green in color; Burnham points out that truly green stars are close companions to red stars (such as the companion to Antares), and *beta Librae* doesn't fit that category. Still, the impression apparently persists for some observers; you'll have to decide for yourself.

Double stars:

Alpha^2 and *alpha^1 Librae* form a very wide double with color contrast: yellow and pale blue. Note that *alpha^2* is the primary: 2.9, 5.3; PA 314, separation 231".

Iota Librae is a multiple system:

> The companion iota^{1a} is a rapid binary with a period of

22.35 years, traveling in a retrograde motion.

Iota^{1B} is a fixed wide companion: 4.5, 9.5; PA 111, separation 58.6".

Struve 1962 is a fixed pair of equal stars: 6.5, 6.6; PA 188 degrees, separation 11.9".

Variable stars:

Delta Librae is an Algol-type variable: 4.9-5.9 with a period of 2.3 days.

48 Librae (also known as *FX Librae*) is a noted shell star that may be dormant for many years, then show rapid activity.

The star has an exceptionally large rotational velocity, and (perhaps as a consequence) an equatorial ring of gases about twice the diameter of the star which rapidly expands.

Deep Sky Objects:

The only notable deep sky object is a rather loose globular cluster of faint stars: NGC 5897, thought to be about 50,000 light years away. The larger the telescope, the better the impression.

The cluster is found two degrees southeast of iota Librae.

Lupus

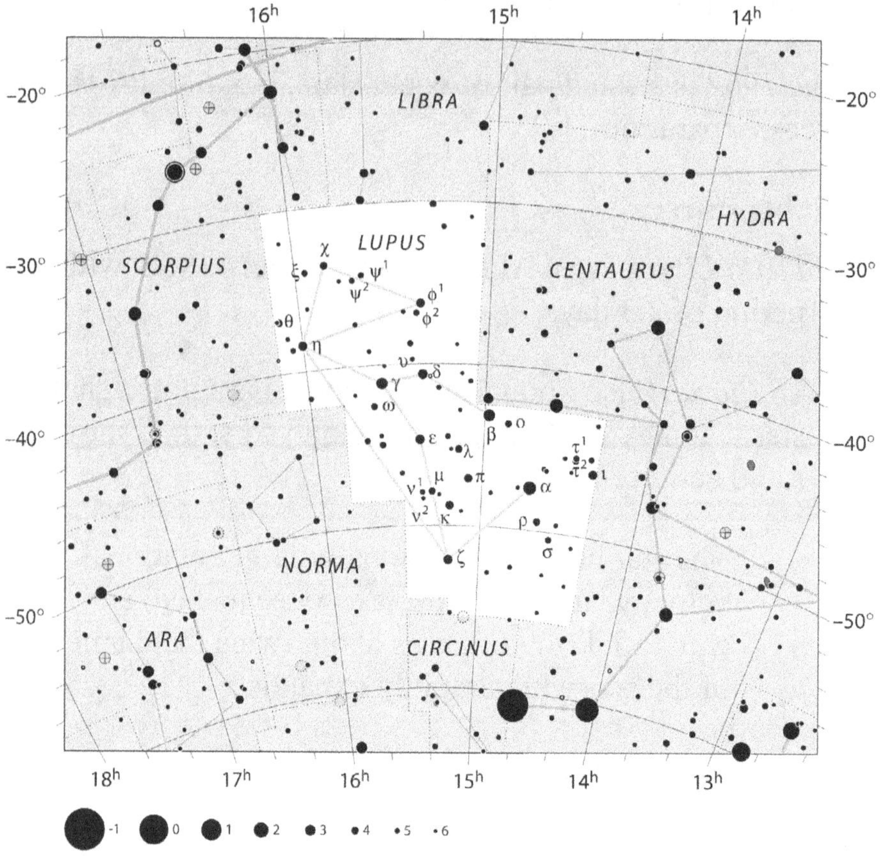

The Wolf

The constellation may refer to the ancient king of Arcadia, King Lycaon (a word which is related to both 'wolf' and 'light'). King Lycaon, who ruled Arcadia with his fifty sons, was said to continue the practice of human sacrifice when other parts of Greece had abolished it as barbaric.

> One variation of the story has King Lycaon offering Zeus the sacrifice of a young boy, which angered Zeus so much he promptly changed Lycaon into a wolf and struck his house down with lightning, killing all his fifty sons.

> In another version Zeus one day visited Arcadia disguised as a simple traveler. Lycaon and his sons offered him soup made not only from the meat of goats and sheep, but also of his own son Nyctimus. Zeus overthrew the table in disgust and killed all the king's sons with lightning bolts (restoring the life of Nyctimus in the process).

> While it may seem ludicrous to memorialize this barbaric King of Arcadia, even prior to the Greeks the constellation seems to have represented a sacrificial animal of some sort.

The asterism is not very obvious, and in fact the constellation itself is fairly difficult to separate from its closest neighbour Centaurus.

Lupus is narrowly squeezed between Centaurus and Hydra to the west, Scorpius and Norma to the east. Although small, since it lies in the Milky Way it is packed with interesting items, especially double stars.

The Bayer stars are generally third and fourth magnitude. To locate Lupus, one might first begin with the arms of Centaurus (if need be, one can review this constellation from the Archives). Starting at *theta Centauri* starhop to *phi Centauri* then *eta* and finally *kappa Centauri*.

In the same field to the south is *beta Lupi*, which at 2.7 magnitude is slightly brighter than *kappa Centauri* (3.1). Now drop down to the southwest some five degrees and you'll find *alpha Lupi*.

Alpha Lupi (a beta Cephei variable) is the brightest star of the constellation at 2.3 magnitude. There is some dispute over its distance; some authorities put it at 430 light years, while others at around 620. From the absolute magnitude of -4.4, it would have a distance of 710 light years.

Double stars:

Gamma Lupi is a very close binary with a nearly edge-on orbit whose period is 147 years. Currently the companion is at the greatest distance: PA 274°, separation 0.68".

Epsilon Lupi is also a close binary: 3.4, 5.5; 247°, 0.6".

Eta Lupi is a pleasant fixed binary with slight colour contrast: 3.4, 7.8; 20°, 15".

Kappa^1 and *kappa^2* form a wide fixed binary: 3.9, 5.7; 144°, 27".

Mu Lupi is a multiple system. AB: 5.1, 5.2; 142 degrees, 1.2". The third component is much easier: PA 130°, 24".

Xi^1 and *xi^2 Lupi* are a fixed pair. This double is the most attractive binary in Lupus: 5.3, 5.8; 49°, 10.4".

Variable stars:

Lupus has three beta Cephei variables among its Bayer stars.

> *Beta Cephei* variables are pulsating variables (prototypes are beta Cephei and beta Canis Majoris). These are massive and very luminous stars which rapidly rotate. The range of magnitude change is very small.

Alpha Lupi: 2.29-2.34; period 6h14m.

Delta Lupi: 3.2-3.24; period 3h58m.

Tau^1 Lupi: 4.54-4.58; period 4h15m.

Deep Sky Objects:

The constellation has no outstanding deep sky objects, with the following clusters being perhaps the best examples.

NGC 5822 is a very large open cluster of about a hundred stars. The cluster is about 6000 light years away, and is located 3° SW of zeta Lupi.

NGC 5986 is a globular cluster about 45,000 light years distant. It is 2.5° WNW of eta Lupi.

Lynx

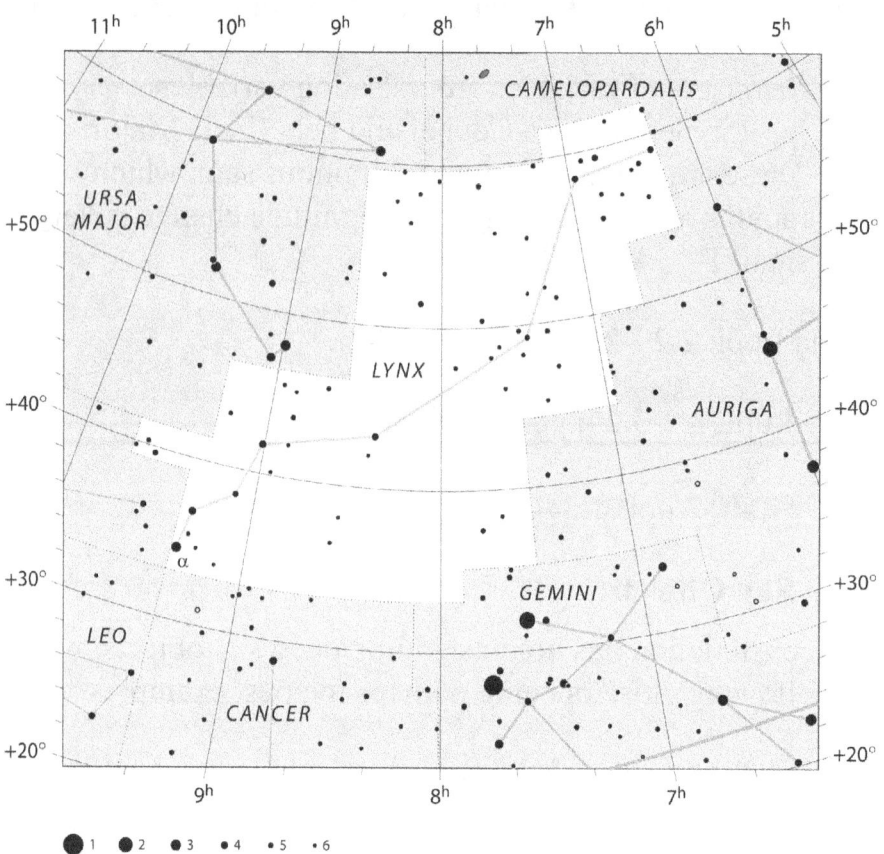

The name *Lynx* never stood for the animal itself. Hevelius, who invented the constellation, said anyone who wanted to study the stars here should have eyes like a lynx.

As with many other minor constellations invented by Hevelius and others to fill in the blanks, Lynx is nothing more than a bumpy line running south from *2 Lyncis* down to *alpha Lyncis*, which sits just north of the border of Leo with Cancer.

> Actually when you connect the stars, the figure of a giant seagull comes to mind, or an albatross perhaps.

The only Bayer star in the constellation is *alpha Lyncis*, a red giant roughly 25-30 times the size of the Sun, and 170 light years away. The star is also a binary, discussed below. In fact, obscure as the constellation is, it has some very fine binaries.

Double stars:

> *12 Lyncis* is about 200 light years away, and the triple system (also known as Struve 948) is an excellent test for telescopes.
>
>> Companion B is 1.7" from the primary, at PA 69 degrees. The orbit is a long one, 699 years, and as you can see it describes a near-perfect circle.
>>
>> Companion C is a fixed star at PA 308, separation 8.7". With a medium sized telescope you should be able to resolve all three components.
>
> *Struve 958* is another wonderful sight: two 6.0 stars, PA 257 degrees, 4.9". To find it, move south four degrees from 12 Lyncis (passing 13 Lyncis on the way).

Struve 1009 is perhaps even more attractive. Find 15 Lyncis, the brightest star in this corner of the constellation. Then drop down five and a half degrees and east about a degree.

After enjoying the delicate beauty of these last two binaries, *19 Lyncis* offers a different kind of sight.

> There's a small challenge here, for some observers report the companion to be purplish; others a soft green. This too is a Struve binary (Struve 1062): 5.3, 6.6; PA 315 degrees, 14.7".

Kui 37 is one binary better known by another name: *10 Ursae Majoris*. With a proper motion of .507" in a westerly direction (240 degrees) this star has moved from Ursa Major to Lynx but kept its old name.

> The star is a close visual binary with the companion revolving its primary every 21.9 years. The Epoch 2000 values are: PA 46 degrees and separation a mere 0.6".

Finally we offer *Struve 1282*, a faint binary system which reflects the same delicate beauty seen in the others listed.

> This system is located in the eastern region, not far from *alpha Lyncis*.

> From *alpha Lyncis* move west six degrees then north nearly one degree. The right ascension is 8h50m44s, declination 35 degrees, 4 minutes. The binary is the brighter of the small group of stars

found here.

Variable Stars:

R Lyncis is a long-period variable with a range from 7.2 to 14.3, and a period of 387.75 days. In 2000 the maximum should occur roughly on the tenth of March.

Deep Sky Objects:

NGC 2683 is a spiral galaxy which lies two degrees to the southeast (or a degree northwest of sigma^2 Cancri). Seen practically edge-on, it's fairly bright and quite large.

NOTES

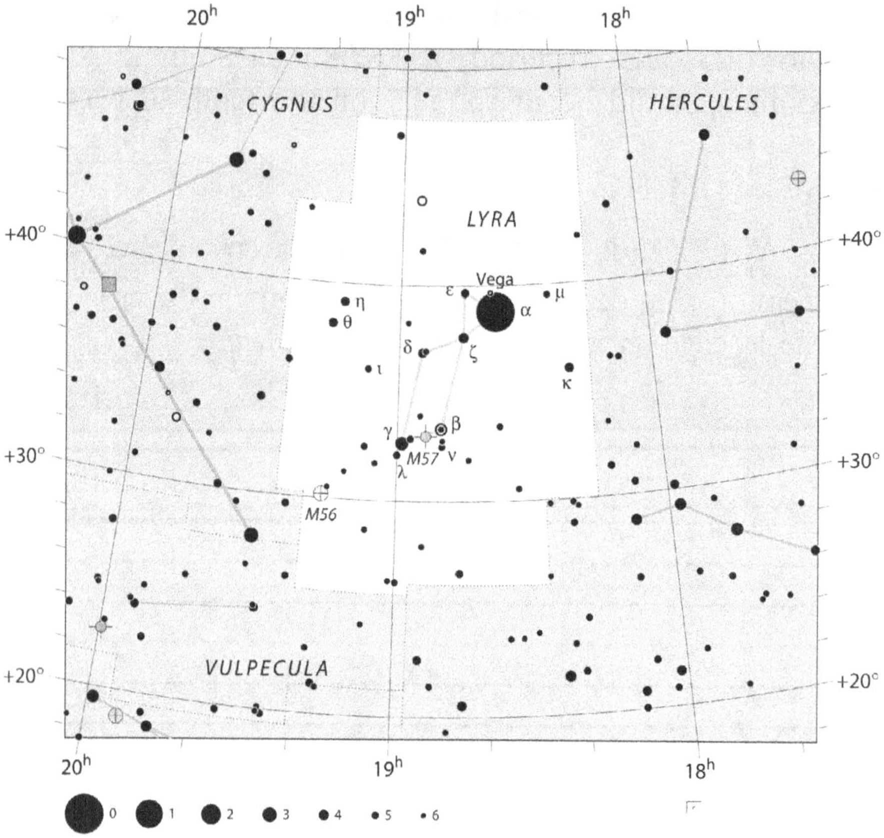

The Lyre

Lyra represents the lyre played by Orpheus, musician of the Argonauts and son of Apollo and the muse Calliope. Apollo gave his son the lyre as a gift, and Orpheus played it so well that even the wild beasts, the rocks, and the trees were charmed by his music. He fell deeply in love with the nymph Eurydice, and the two were married. Their wedded bliss did not last for very long, however. Eurydice was wandering in the fields with some other nymphs when she was seen by the shepherd Aristaeus. Aristaeus was struck by her beauty and pursued her; as she fled, she was bitten by a snake in the grass and died of the serpent's poison.

Orpheus was devastated. He decided to seek out his wife in the underworld, and gained an audience with Pluto and Persephone. The king and queen of the underworld, like all others, were charmed by his music and granted him permission to take Eurydice back to the land of the living with him:

They called Eurydice. She was among the ghosts who had but newly come, and walked slowly because of her injury. Thracian Orpheus received her, but on condition that he must not look back until he had emerged from the valleys of Avernus or else the gift he had been given would be taken from him.

Up the sloping path, through the mute silence they made their way, up the steep dark track, wrapped in impenetrable gloom, till they had almost reached the surface of the earth. Here, anxious in case his wife's strength be failing and eager to see her, the lover looked behind him, and straightaway Eurydice slipped back into the depths. Orpheus stretched out his arms, straining to clasp her and be clasped; but the hapless man touched nothing but yielding air. Eurydice, dying now a second time, uttered no complaint against her husband. What was there to complain of, that she had been loved? With a last farewell which scarcely reached his ears, she fell back again into the same place from which she had come (Metamorphoses X 47-63).

According to Ovid, Orpheus was so heartbroken from having lost his love not once, but twice, that he rejected the company of women in favor of that of small boys. The women of Thrace were infuriated and, while maddened during Bacchic rites, hurled rocks at the bard. The rocks, tamed by the sound of Orpheus's lyre, at first fell harmlessly at his feet, but the shrieks of the infuriated women soon drowned out the music. The women dismembered Orpheus, throwing his lyre and his head into the river Hebrus. The Muses gathered up his limbs and buried them, and Orpheus went to the underworld to spend eternity with Eurydice. Zeus himself cast the bard's lyre into the sky.

The constellation is small and rather faint, but it is home to the fifth brightest star, Vega. The asterism resembles some multi-legged creature more than it does a lyre, with Vega at the head.

The constellation hasn't the full complement of Bayer stars, and only three stars are brighter than fourth-magnitude. Still, there are some very fine objects to view.

Vega, "Falling Eagle" or "The Harp Star", is only the fifth brightest star, but it dominates the summer skies in the northern hemisphere, with a transit date of 1 July.

About 12,000 years ago Vega served as the Pole Star, and it will again in another 12,000 years.

Beta Lyrae, sometimes known as "Sheliak" (Tortoise), is a prototype of a variable star in which a close companion is transferring matter to its gigantic primary. In Beta Lyrae's case, the transference is occurring very rapidly. Eventually the system will become an Algol variable. (See below for its values.)

Double stars:

Delta^2-Delta^1 Lyrae form a wide binary that may be gravitationally bound despite the great distance. The two have a nice colour contrast, orange and blue. Note that delta^2 is the primary: 4.3, while delta^1 has a visual magnitude of 5.6.

Beta Lyrae is a fixed multiple binary, with a primary of 3.5. AB: 3.5. 8.6; 149°, 46"; AE: 9.9, 318°, 67"; AF: 9.9, 19°, 85".

Epsilon^1-Epsilon^2 Lyrae: the famous "Double-Double". All four stars are fifth-magnitude. The two principal stars form a very wide binary: PA 173°, separation 208".

Each star is itself a double:

Epsilon^{1A}-Epsilon^{1B} is a slow binary with 1165 year orbit: 5.0, 6.1; PA 350° and separation 2.6".

Epsilon^{2C}-Epsilon^{2D} is about twice as fast, with a period of 585 years: 5.2, 5.5; PA 83.5°, separation 2.3".

Zeta Lyrae is another relatively fixed multiple. The brightest component is D: AD 4.3, 5.9; 150°, 43.7". The other components are fourteen magnitude.

Struve 2470 and *Struve 2474* form another fine double-double, that some say is equal to epsilon[1,2].

Struve 2470: 6.6, 8.6; 271 degrees, 13.4" and *Struve 2474*: 6.5, 8.6; 261°, 16.4".

The two binaries are found two and a half degrees NE of gamma Lyrae, which is the brightest star in the region. Or, if you can find iota Lyrae, drop south one and a half degrees. It's a sight well worth the detour!

Variable stars:

Alpha Lyrae (Vega) is a delta Scuti type variable, fluctuating from -0.02 to 0.07 every four hours 33.6 minutes.

Beta Lyrae is an EB variable: 3.25 to 4.4 with a period of 12h22m.

RR Lyrae is a prototype for a pulsating type of variable with short periods, usually less than twenty-four hours. RR Lyrae's period is 13h36m, and it changes in magnitude from 7.1 to 8.1.

Deep Sky Objects:

There are two Messier objects in Lyra: M56 and M57.

M56 (NGC 6729) is a globular cluster, very condensed. It is found eight degrees due south of theta Lyrae.

M57 (NGC 6720) known as the *Ring Nebula*, is the finest planetary nebula in the skies. The ring itself should be clearly visible in medium scopes, while the fourteen magnitude central star may take a little longer. Burnham gives an excellent discussion on this object.

It is located between beta and gamma Lyrae (slightly closer to beta), and is about 4000 light years distant.

NOTES

Mensa

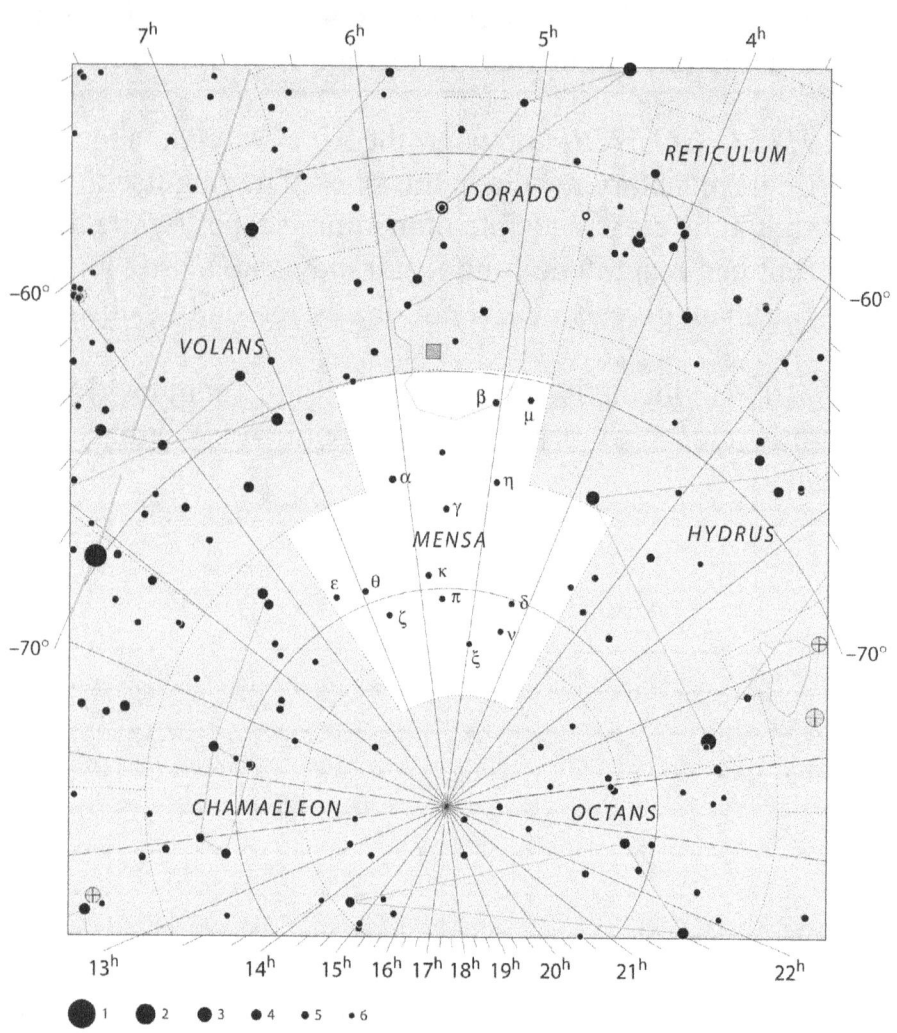

The Table

Mensa, *The Table*, is another of Nicolas de Lacaille's creations, this one named for the Table Mountain at the Cape of Good Hope, where Lacaille observed the southern hemisphere skies in the mid eighteenth century.

The asterism shows an upside down mountain top. The mountain is seen right-side up in the Southern Hemisphere around midnight in mid-July.

There are a little over a dozen Bayer stars in Mensa, mostly fifth magnitude.

Double stars:

h3607 is a binary with colour contrast, yellow and blue: 8, 8.5; PA 127°, separation 37.2".

The binary is 2° WSW of delta Mensae.

Variable stars:

U Mensae is a semi-regular, 7-10, with a period of approximately every 410 days. It's found a half degree southwest of nu Mensae.

Deep Sky Objects:

The Large Megallanic Cloud is just above the (upside-down) mountain. Beta Mensae is inside this vast galaxy, which is a satellite of the Milky Way, but most of the LMC is in Dorado.

Microscopium

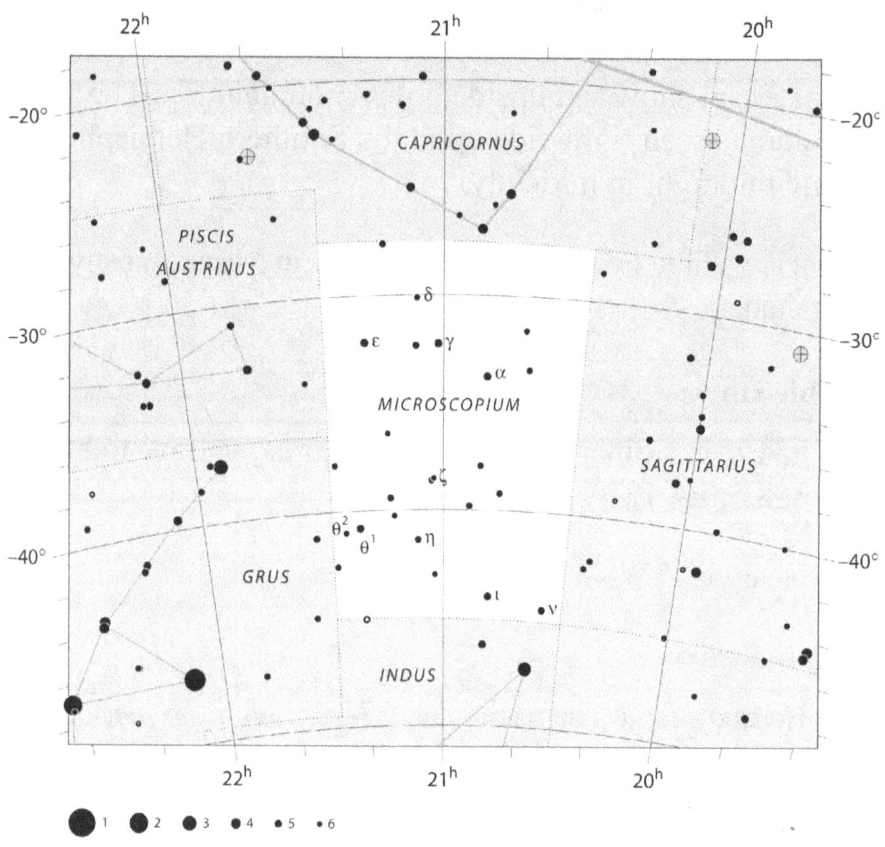

The Microscope

Microscopium is a small constellation in the southern sky, defined in the 18th century by Niholas Louis deLaxaille. Its name is Latin for microscope, it was called this due to its visual similarity to the 18th century microscope. Its stars are very faint and hardly visible from most of the non-tropical norther hemisphere.

The former constellation Neper, representing an auger, may have been located in or near modern-day Microscopium. However, this connection is disputed.

Double stars:

Alpha Microscopii is a fixed binary with faint companion: 5.0, 10.6; PA 166, separation 20.5".

Theta^2 Microscopii: 6.4, 7.0; PA 267, 0.5".

Variable stars:

Theta^1 Microscopii is an alpha CV type star of small range: 4.77 to 4.87 every 2d2h55

Deep Sky Objects:

There are no deep sky objects of interest to amateur observers.

Monoceros

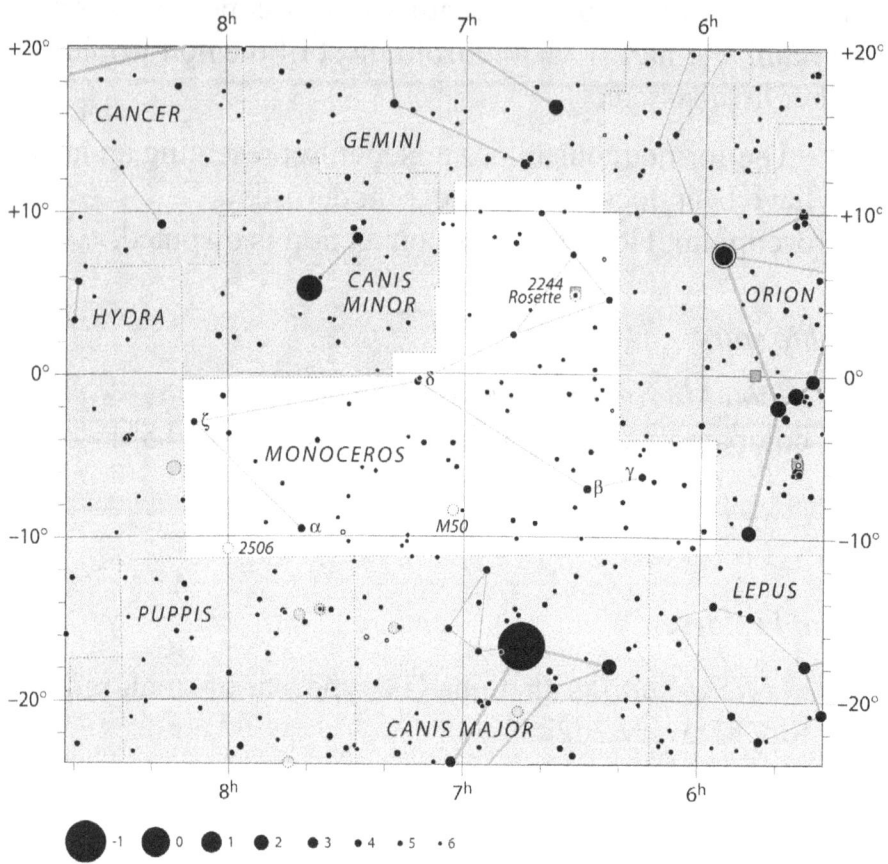

The Unicorn

Flanked by Orion and Canis Minor, with Gemini above and Canis Major below, the faint constellation Monoceros ("the Unicorn") is often overlooked.

While the constellation may have been in existence prior to the seventeenth century, its first historical reference appears in Jakob Bartsch's star chart of 1624, under the name "Unicornu". It is believed that Bartsch (who incidentally was Johannes Kepler's son-in-law) relied on earlier works, but such works have never been identified.

It takes a lot of imagination to fashion a unicorn out of this group of stars. In fact, there are several variations.

But it isn't the stars which hold most of our interest here. Instead, Monoceros has several celebrated deep sky objects as well as the most massive binary system yet discovered.

The stars of Monoceros are as dim as the constellation's history: only a few fourth-magnitude stars that are difficult to notice except on very clear nights. *Alpha Monocerotis* is only a 3.93 visual magnitude, slightly brighter than *gamma Monocerotis*: 3.98.

The only Bayer star of interest is *beta Monocerotis*, which is a splendid triple. See below for a detailed description of this system.

Double stars*:*

Beta Monocerotis is a wonderful triple star system, especially for smaller telescopes. William Herschel, who discovered it in 1781, thought it was one of the best he'd ever seen.

> The three stars form an elegant triangle that doesn't change much, if at all, over time. Thus the system may be considered "fixed". The visual magnitudes and separations are as follows: AB (4.7, 5.2; 132 degrees, 7.3"), AC: (6.1, 124 degrees, 10").

Epsilon Monocerotis is a fixed binary: 4.5, 6.5; 27 degrees, 13".

15 Monocerotis, also known as *S Monocerotis*, is another multiple system consisting of six stars. However most of them are extremely faint:

> **AB: 4.8, 7.6; 213 degrees, 2.8". Component C: 9.9, 13 degrees, 17"; D: 9.7, 308 degrees, 41", E: 10, 139 degrees, 74", and F: 7.8, 222 degrees, 56".**

Variable stars:

R Monocerotis is an irregular variable, and the nucleus of *Hubble's Variable Nebula* (see below). It is an RW Aurigae type variable, changing from visual magnitude 10 to 12.

S Monocerotis is also irregular, the central star in NGC 2264. This star is a bright 4.5 visual magnitude, dipping down at times to about 5.0.

Deep Sky Objects:

NGC 2237, a large diffuse nebula ("Rosette Nebula") which engulfs the open star cluster NGC 2244 (see below). This nebula actually carries four separate NGC numbers (2237, 2238, 2239, and 2246) although it usually goes under the name of NGC 2237.

> It takes a large telescope to distinguish the ring shape. Usually all one sees is a ghostly bit of fluff around the star cluster.
>
> This nebula has been extensively studied, for it seems to be extraordinarily massive (over 10,000 Suns). Dark matter is woven in and out of the surrounding gases. It is surmised that eventually the gases will coalesce, producing either a new star or perhaps even a whole new system of sun and planets, similar to our own.

NGC 2244, the open cluster at the centre of the Rosette Nebula, may actually be stars formed out of the Rosette Nebula. However, the central star, 12 Mon (magnitude 6), probably does not belong to the group.

NGC 2264 is a large and bright cluster with associated nebula (The Cone Nebula, so called because of its shape). The brightest star here is the variable S Monocerotis, which is found near the top of the cluster.

> Like other clusters in this constellation, NGC 2264 is surrounded by gaseous matter not revealed in small scopes. The spectacular dark Cone Nebula is found at the southern edge of this cluster. However, much like the Horsehead Nebula, it appears best in long-exposure photographs.

M50 (NGC 2323) is surprisingly the only Messier object in this constellation.

> This is a cluster of about a hundred bright stars, rather tightly grouped, ideal for small telescopes. It can even be seen by the naked eye on a good night. There is a red star near its centre. The cluster is considered to be about 2500 light years away.
>
> To find M50 draw a line between Sirius and Procyon; you'll find the cluster about a third of the way up from Sirius.
>
> Another way to find M50 is to locate the roughly-shaped square formed by alpha, delta, and beta Monocerotis, along with Sirius. Right in the middle of that square lies M50.

Then there is Plaskett's Star:

This giant double star system is recognized as the most massive pair yet discovered.

> John Stanley Plaskett began his career at the Dominion Observatory in Ottawa. As he became aware of its limitations, he lobbied the Canadian government to support the development of a new astronomy facility.
>
> In 1913 the federal government provided funding for the construction of a 183 cm (72 inch) reflecting telescope, to be built near Victoria, BC. The Dominion Astrophysical Observatory officially opened in 1918 and was, for a time, the largest telescope in the world. Plaskett served as DAO's first director from 1917 to

1935.

It was here that Plaskett set about studying binary stars and in 1922 this work resulted in his discovering the very massive binary star which now bears his name.

The system is comprised of two giant O-type stars, each of which orbits a common center of gravity every 14.4 days. While Plaskett arrived at a mass of 90 Suns for each star, it is now probable that the total mass of the two does not exceed 100 Suns. Even so, this pair still stands as the most massive double star sytem yet discovered.

Plaskett's Star is probably a member of the NGC 2244 cluster (see above).

Hubble's Variable Nebula (NGC 2261) and the mystery star R Monocerotis

The nebula has a curious shape, somewhat like a comet's tail. At the "head" of the comet is where the variable R Monocerotis will be found ... perhaps.

This reflection nebula has a usual visual magnitude of about 10, but this fluctuates sporadically. It was originally thought that as R Monocerotis's visual magnitude changed, so did the visual magnitude of the nebula. But this proved to be false; the nebula's variations do not seem to be associated with the star's variability.

However, Valerie Illingworth *(Facts On File Dictionary of Astronomy)* states that the variability of the nebula comes from what is called a bipolar flow of emissions originating from R Monocerotis. This ejection of gas in two opposite directions is typical in very young stars.

Other observers question the existence of R Monocerotis, considering the area nothing more than an extremely dense gaseous area. Tirion's *SkyAtlas 2000.0* doesn't show the star, and major star catalogues don't list the star. Burnham calls it "a bright nebulous condensation with perceptible apparent size."

Observations at Kitt Peak and Mauna Kea have concluded that R Monocerotis is a protoplanetary system. That is, that planets may presently be forming in a highly condensed region: another "solar system" being born.

While the nebula is easily seen in small scopes, it is a little tricky to find. *Burnham's Celestial Handbook* has a finder's guide (p. 1201). Or you may try this: once you locate epsilon Monocerotis star-hop up to 13 Monocerotis. Farther up, to the northeast, is S Monocerotis. Between these two, just about half way, the great nebulosity surrounding S Monocerotis begins. At the extreme southern edge of this nebulosity is the distinguishing form of Hubble's Variable Nebula: the comet-like shape is unmistakeable.

NOTES

Musca

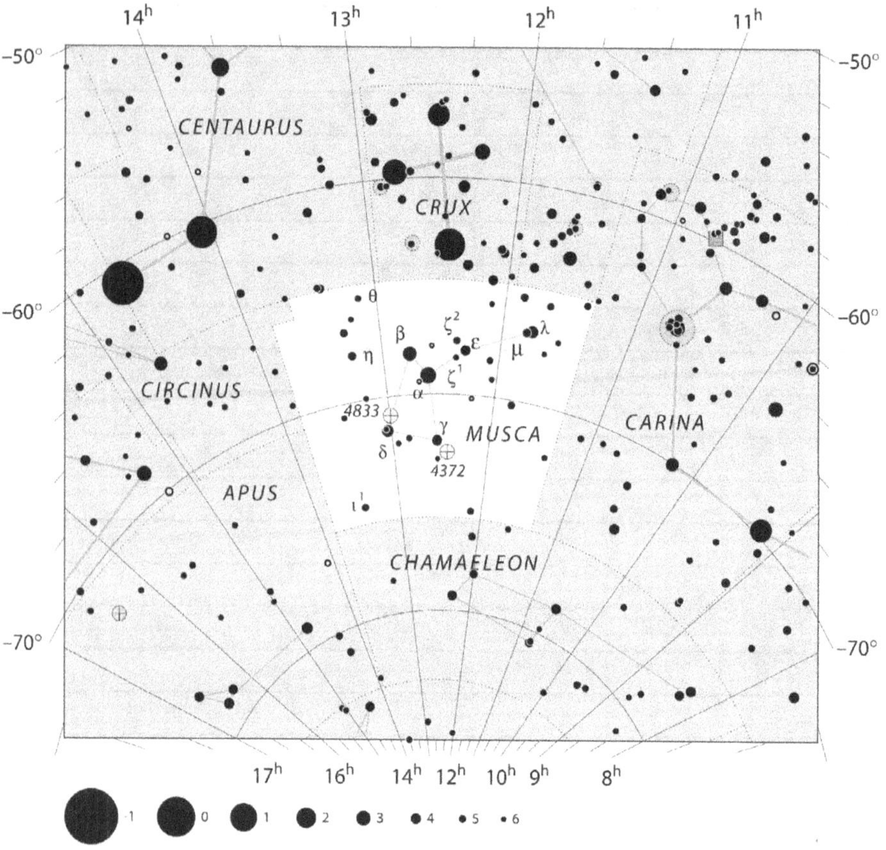

The Fly

Musca, "The Fly", is a southern hemisphere constellation introduced by Johann Bayer. He called it Apis, "The Bee"; perhaps because of its similarity with "Apus", this name didn't stick.

About a third of the Coal Sack Nebula spills over into Musca; most of it is found in neighbouring Crux.

Double stars:

Beta Muscae is a rapid visual binary; the companion circles the primary every 383.12 years: 3.7, 4.0; PA 43 degrees, separation 1.3". 3.5".

Theta Muscae is a fixed binary: 5.5, 8.0; PA 186 degrees, 5.3".

Variable stars:

Most variables here offer very small changes in magnitude, using alpha Muscae as an example. R Muscae is not a long-period Mira, as is usually the case with "R" stars, but rather a cepheid.

Alpha Muscae is a beta Cas type variable: 2.68 to 2.73.

R Muscae is a cepheid varying from 6.4 to 7.3 every 7h30m36s.

Deep Sky Objects:

NGC 4372 is a rather faint globular cluster one degree SW of gamma Muscae.

Norma

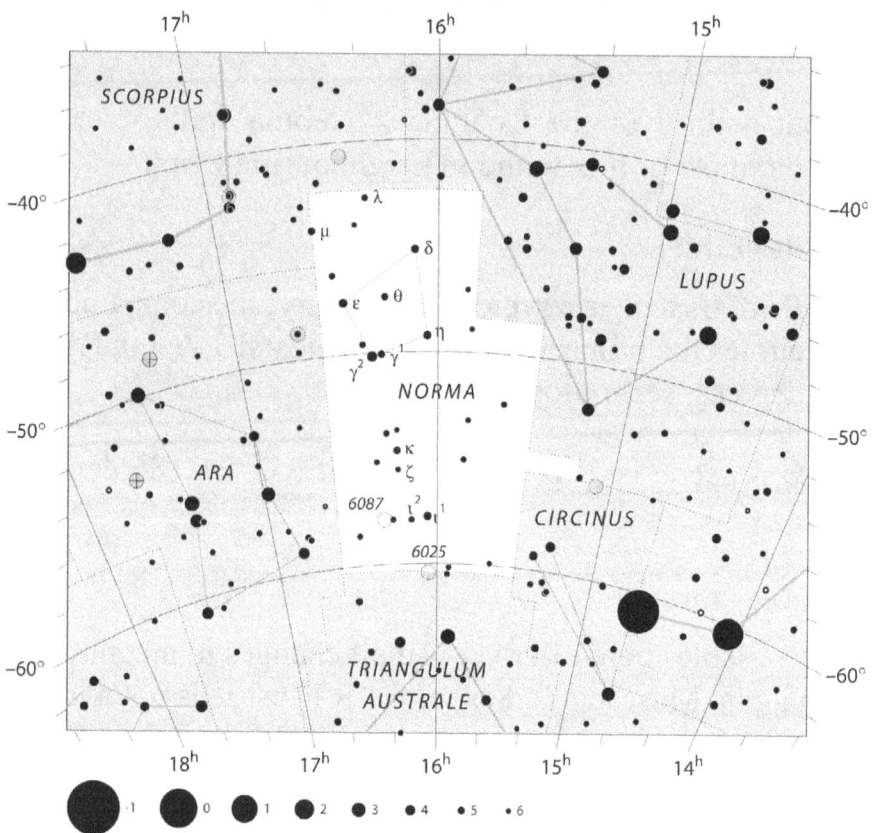

The Level

Norma is another of those relatively insignificant constellations in the Southern Hemisphere. Invented by Nicolas Louis de Lacaille in the mid-eighteenth century, the constellation represents a scientific instrument, "the level". The original name was "Norma et Regula" (the level and the square).

Since Lacaille's time, the principle stars of Norma have been redistributed, leaving it with fewer .

Double stars:

Epsilon Normae is a fixed binary: 4.8, 7.5; PA 335, separation 22.8". Both of these stars also have a spectroscopic companion.

Iota^1 Normae is a multiple. AB is a rapid binary with an orbit of 26.9 years. The Epoch 2000 values are 5.6, 5.8; PA 285, and separation 0.5".

Component C: 8, PA 246, separation 10.8".

Variable stars:

Norma has three variables of (possible) interest:

Mu Normae is an alpha Cygni type variable: 4.87-4.98.

R Normae is a Mira-type variable: 6.5-13.9, with a period of 507.5 days. The next maximum is scheduled for the last week of February 1997; in 2001 the maximum should occur in the last week of April.

S Normae is a well-known cepheid with a range from

6.1 to 6.8 magnitude, every 9.75 days. It is found in the NGC 6087 cluster (see below).

Deep Sky Objects:

Norma has several deep sky objects, including a notable planetary nebula.

> *NGC 6067* is a cluster of about a hundred tenth-magnitude stars. This cluster is in the same field as kappa Normae, just to the north of this star.
>
> *NGC 6087* is another cluster, some 3500 light years away, comprised of forty or so stars, ranging from 7-10 magnitude. The group is two degrees east of iota[1]; it includes the cepheid variable S Normae.
>
> *Sp 1* is a planetary nebula, rather bright and perfectly circular with a 13 magnitude star in its centre. The planetary nebula is five degrees west-southwest of gamma[2] Normae.

NOTES

Octans

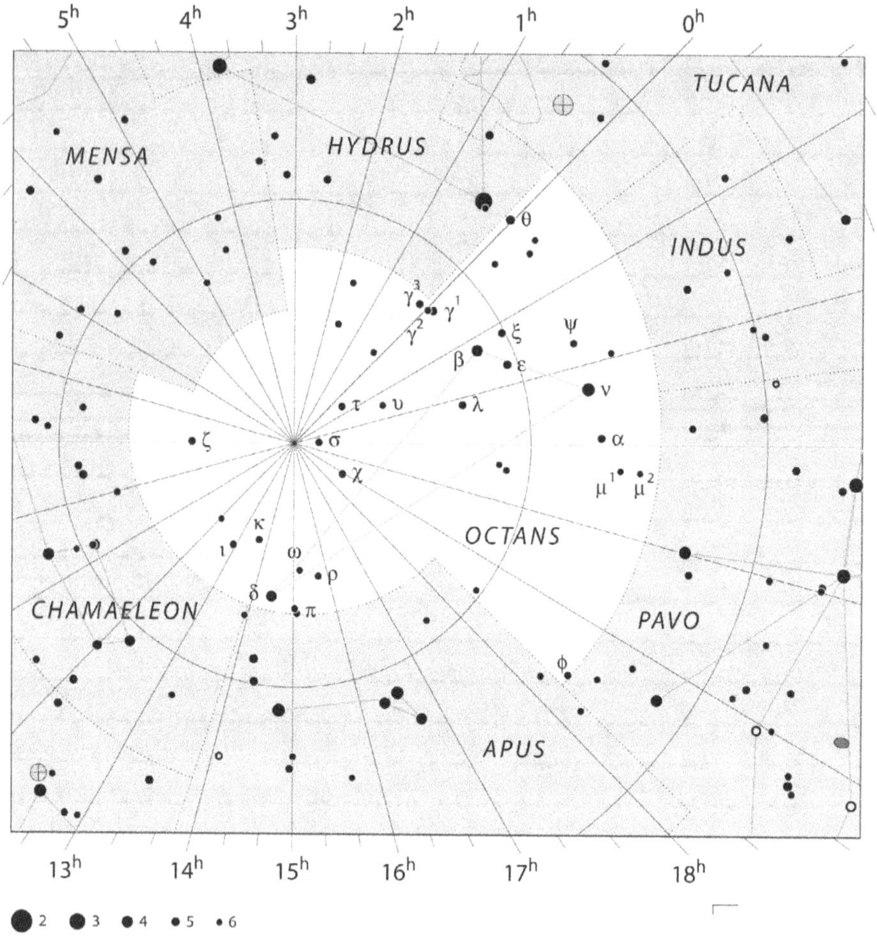

The Octant

Octans is a rather non-descript constellation which includes the southern polar region. The constellation was devised by Nicolas Louis de Lacaille in 1752. It commemorates the octant, which was invented by John Hadley in 1730. In fact the full name of the constellation is "Octans Hadleianus".

The instrument divided the circle into eight parts, which facilitated the making of angular measurements in both astronomy and navigation.

Most of Octans' stars are fifth magnitude, including *sigma Octantis*, the Southern Pole Star (which is actually about a degree from the true south pole and moving slowly away from it).

Rather strangely, the brightest star in the constellation is *nu Octantis,* an orange giant 64 light years away.

Octans has one notable binary, a few variables, and little else of interest.

Double stars:

Lambda Oct is a double with eighth magnitude companion: 5.5, 7.8; PA 70 degrees, separation 3.1".

Variable stars:

Epsilon Oct is a semi-regular variable: 4.6 - 5.3 every 55 days or so.

Sigma Oct is a delta Scuti variable, 5.45 - 5.5 every 2h 19m 41s.

Ophiuchus

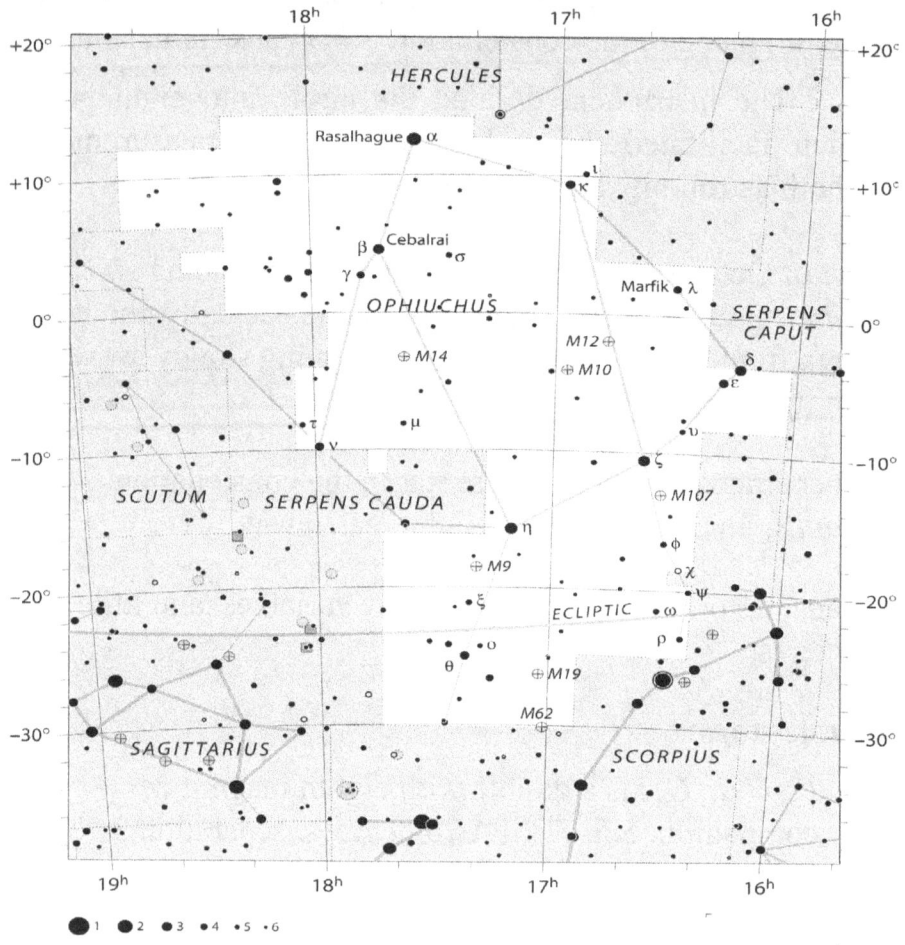

Serpent-Bearer

There is some disagreement over the origin of this constellation. Apparently it was once known as Asclepius, who was the Greek god of medicine. One such reference was made in the writings of Eudoxus, in the fourth century BC.

Eudoxus (c400-c347 BC) deserves to be better known. He may have been a member of Plato's Academy, and it is possible he was its head for some time.

Eudoxus was a prolific writer of scientific subjects, and thinkers such as Euclid incorporated much of his work into their own. He mapped out the constellations, and the result became the main star reference for hundreds of years. Among other feats, he divided the sky into degrees of longitude and latitude and devised a better calendar. He was also a well known geographer and mathematician, but it was his work on astronomy that he is principally remembered.

Later Greek stories arose about Carnabon, a king of the Getae, who killed a famous dragon, or even of Heracles, who (you might recall) killed Draco. Thus the story of the man and serpent came to represent a host of individuals, but most authorities now seem to opt for Asclepius, or Aesculapius, which is the Latin equivalent of the Greek god of medicine.

Ancient sculpture typically shows the god bare breasted, attired in a long flowing cloak, and holding a staff with a serpent coiled about it. This is perhaps the forerunner of the modern medical symbol of the caduceus.

The constellation Ophiuchus is thus found in the midst of the Serpens. The southern part of Ophiuchus dips into a very dense portion of the Milky Way, resulting in a great many deep sky objects.

The Bayer stars of Ophiuchus are fairly bright, five of which have a magnitude brighter than 3.0.

The brightest star, alpha Ophiuchi, is better known as *Rasalhague*, meaning "Head of the Snake Charmer". This is a rather close star, at 54 light years away, and a celestial neighbour of Ras Algethi (alpha Herculis), which lies to the WNW five degrees.

> Ophiuchus has a half-dozen or so visual doubles, and even more star clusters. In fact Ophiuchus has more globular clusters than any other constellation.

The region encircling rho Ophiuchi is also of some interest. This area contains several dark clouds and nebulae that show the active formation of stars.

Double stars:

Ophiuchus has one of the finest collections of double stars, including several close visual binaries.

> *Eta Ophiuchi* is a close visual with an orbit of 88 years: 2.9, 3.4; presently the companion is at PA 247° and separation 0.6".
>
> *Lambda Ophiuchi* is also a rapid binary. 4.2, 5.2; currently the PA is 27° and its separation is 1.5".
>
> *Xi Ophiuchi*: 4.5, 9.0; PA 50°, separation 3.7".

Rho Ophiuchi: 5.3, 6.0; PA 344°, 3.1".

Tau Ophiuchi: 5.2, 5.9; with an orbit of 280 years. Presently the companion is at PA 282° and separation 1.7".

36 Ophiuchi is a binary with period of 548 years, of two equal stars: 5.1, 5.1; 148°, 4.9".

70 Ophiuchi is another close binary with a period of 88.3 years. 4.2, 6.0. In 2000.0 the values are PA 149° and separation 3.7".

Struve 2276. This is a very beautiful fixed binary of two fairly faint stars: 7.0, 7.4; PA 257°, separation 6.9".

Variable stars:

Kappa Ophiuchi is an irregular (Lb) variable that fluctuates betweem 4.1 and 5.0.

Chi Ophiuchi is a gamma Cas variable: 4.2-5.0.

U Ophiuchi is an Algol type (EA) variable: 5.84-6.6 every 1.7 days.

X Ophiuchi is a long-period variable, 5.9-9.2 with a period of 328.85 days. In the year 2000 the maximum should occur during the last week of March.

Deep Sky Objects:

There are six Messier objects in Ophiuchus: M9, M10, M12, M14, M19, and M62 (and one more as well, if you accept M107 as a true Messier). These are all globular clusters.

M9 (NGC 6333) is the smallest of this group, unresolved except in large instruments.

> The cluster is found 3.5 degrees SE of eta Ophiuchi. It is considered to be about 26,000 light years away.
>
> In the same field are two more globular clusters: NGC 6342 (1 degree SE) and NGC 6356 (1 degree NE).

M10 (NGC 6254) and *M12 (NGC 6218)* are nearly identical globular clusters: like tiny explosions of stars with dense cores.

> M12 is eight degrees north of zeta Ophiuchi and two degrees east. M10 is 2.5 degrees SE of M12, with 30 Ophiuchi in the same field.

M14 (NGC 6402) needs a 20-cm telescope to resolve; it's more condensed than the preceding two and slightly fainter.

M19 (NGC 6273) is another very dense cluster, usually described as "oblate", meaning it's a bit egg-shaped. It is about 25000 light years away.

> M19 is seven degrees due east of Antares (alpha Sco), or two and a half degrees west of the bright double 36 Ophiuchi (and very slightly north, less than a degree).

M62 (NGC 6266) is six degrees SW of theta Oph (and four degrees south of M19); this is another non-circular globular cluster, a little brighter than M19. (Note:

Burnham includes this Messier in Scorpius; nearly all other authorities put it in Ophiuchus.)

M107 (NGC 6171) is the faintest of the bunch and quite small. This is one of those "Messiers" that were added to the original list, for some reason. It's three degrees SSW of zeta Ophiuchi.

B78, the "Pipe Nebula", is a naked eye dark nebula two degrees southeast of theta Ophiuchi, in very rich area of the Milky Way.

Barnard's Star is the most rapidly moving star relative to the solar system, with a proper motion of 10.31", and the second closest star to us, at a distance of 5.91 light years (if you accept the three-star system of alpha Centauri as a unit).

This is a red dwarf, with a visual magnitude of only 9.5, and consequently not easily found. Burnham has a finder's chart, page 1253, but since that chart was published the star has moved north 1.1 centimetres.

The star is three degrees due east of beta Ophiuchus. The actual location (Epoch 2000) is R.A. 17h58m; Decl. +04 degrees, 34 minutes.

A slight oscillation in both the right ascension and declination of Barnard's Star has led observers to suggest the possibility that one or more planets orbit the star.

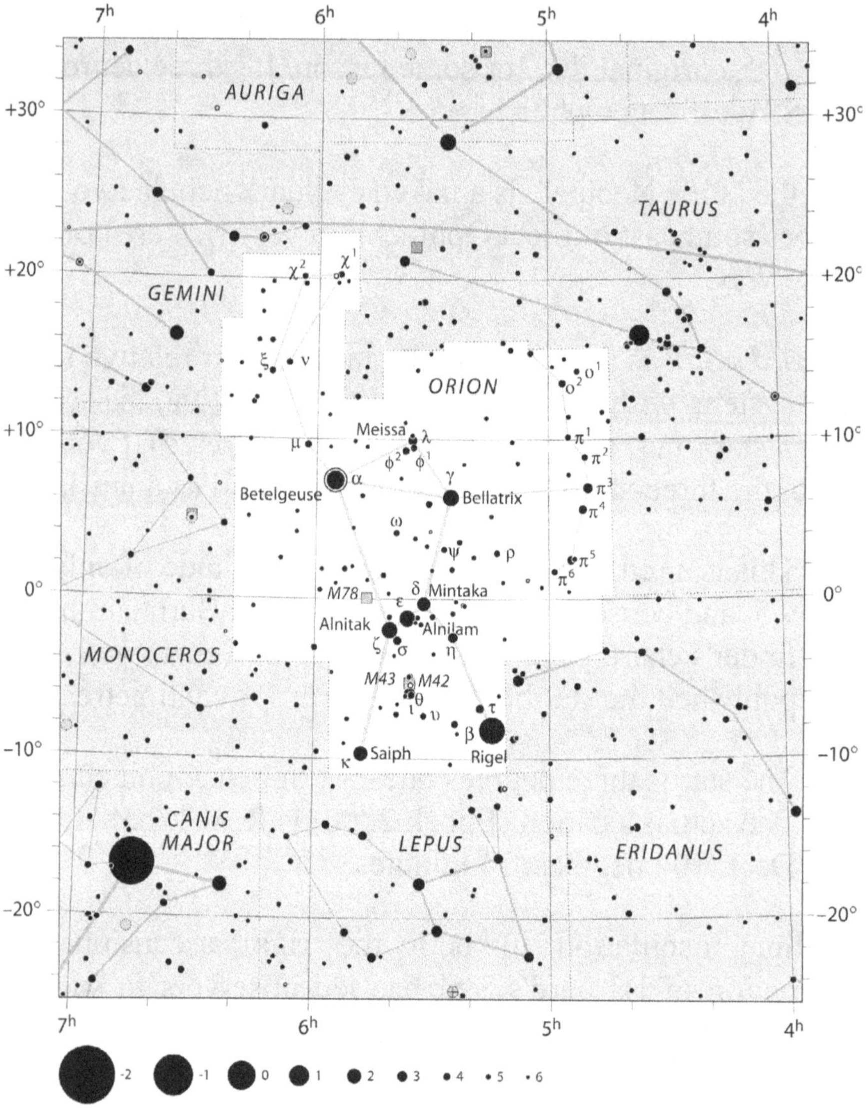

The Hunter

Orion is the master of the winter skies. He lords over the heavens from late fall to early spring, with his hunting dog Sirius trailing at his feet.

The mythic tales of Orion go as far back as the Hittites, who flourished from the Second Millenium BC to around 1200 BC.

> One story from this culture gives an interesting account of Orion's death. Here he is called Aqhat, and was a handsome and famous hunter. The Battle-Goddess Anat fell in love with Aqhat, but when he refused to lend her his bow, she sent another man to steal it. This chap bungled the job, and wound up killing Aqhat and dropping the bow into the sea. This is said to explain the astronomical fact that Orion and the Bow (an older version of the constellation) drops below the horizon for two months every spring.

Like all myths borrowed from several sources over a great length of time, the Greek stories offer many variations. Generally speaking, Orion was known as the "dweller of the mountain", and was famous for his prowess both as a hunter and as a lover. But when he boasted that he would eventually rid the earth of all the wild animals, his doom may have been sealed.

> It might have been the Earth Goddess herself who sent the deadly scorpion to Orion. Or possibly Apollo, concerned that Orion had designs on his sister, Artemis. Thus Apollo may have told the Earth Goddess of Orion's boast. In any case, it seems clear that it was the Earth Goddess who sent the scorpion on its mission.
>
> Some stories have the scorpion killing Orion with its

sting. However the general consensus is that he engaged the scorpion in battle but quickly realised its armour was impervious to any mortal's attack. Orion then jumped into the sea and swam toward Delos. But Apollo had witnessed Orion's struggle with the scorpion and would not let him escape so easily. He challenged his sister Artemis, who was an excellent shot, if she could hit that small black object far away in the sea, the head -- he told her -- of an infamous and treacherous villan. Artemis struck the object with her first shot. She then swam out to retrieve her victim's corpse, and discovered she had killed Orion. Artemis implored the gods to restore his life, but Zeus objected. So she put Orion's image in the heavens.

In his eternal hunting, Orion is careful to keep well ahead of the scorpion. In fact Orion has disappeared over the horizon by the time Scorpio rises in the east, as it becomes his turn to rule the evening sky.

Finding Orion should be no problem. Its stars are some of the most familiar in all the heavens. Question: can you name the three stars that make up Orion's Belt. (Answer below.)

Above the belt, slightly to the left, is *Betelgeuse, alpha Orionis*.

> *Betelgeuse*, the right arm of Orion (or "armpit" as the name suggests), glows with a dull red. Although labelled *alpha Orionis*, it is less bright than *beta Orionis* (Rigel), in the opposite corner of the constellation, to the southwest. Yet if slightly less bright, it is much larger, estimated at around 250 Suns. If one were to replace our Sun with Betelgeuse, its size

would completely engulf the Earth and extend as far as Mars.

As the brightest star in Orion, *Rigel* ranks as the seventh brightest star in all the heavens, just behind Capella. It is a visual binary; its companion is much fainter, but quite visible if you are persistent enough (PA 202°, 9.4").

The other corners of the constellation are formed by *Bellatrix (gamma Orionis)* and *Saiph (kappa Orionis)*. It was once thought that all women born under the sign of Bellatrix would be fortunate and have the gift of speech. The star's name is often translated as Female Warrior or Amazon, and another name sometimes seen is "Amazon Star".

The constellation's main feature is of course the three stars which form the "belt" across the middle of Orion: from west to east *Mintaka, Alnilam,* and *Alnitak*. Even the Bible makes reference to this famous group. God, while pointing out how all-powerful he was, is purported to have asked Job if he (Job) was able to "loose the bands of Orion" (*Job* 38.31).

The last of these stars is also known as *zeta Orionis*, and is a well known triple star system. The primary is a blue-white star, and its companion (165°, 2.3") is a dull red. Close by, just to the south, is the renowned Horsehead Nebula, a so-called dark nebula that is not visible in scopes but quite spectacular in long-exposure photographs.

Binary stars:

There are many double stars in this constellation visible in small telescopes. Below are several selected from a wide list.

Beta Orionis (Rigel) has a 10.4 visual magnitude companion at 202° and a wide 9.5" separation. This is a fixed system.

Lambda Orionis (between Betelgeuse and Bellatrix) is another fixed binary, with a 5.5 companion at PA 43° and 4.4" away.

Theta^1 is a complex system of fixed stars. The four brightest form The Trapezium, an outstanding multiple system for small telescopes. AB is at a position angle of 32° and separation 8.8", AC: PA 132°, 12.7", and AD: PA 96°, 21.5".

Theta^2 is also a fine binary, a triple system to the southeast of The Trapezium. Component B is a binocular object: 6.4 magnitude at a position angle of 92° and separation 52.5". Component C (8.5) is even wider: PA 98° and separation 128.72".

Sigma Orionis is one of the few orbiting binaries found in Orion. Component B has an orbit of 158 years and is one of the few components that traces a not-quite-perfect circle. That's to say, we see it nearly face on, as a wheel spinning around its hub.

The separation never changes much from its current distance of only 0.2". Its 2000.0 position angle is 132°.

Much easier to resolve is component E, with a visual magnitude of 6.7, this is a binocular object at a position angle of 61° and separtion of 42".

Zeta Orionis (1.9, 4.0) has a very slow orbit of 1509 years, and is currently at 165° and 2.3" separation.

Variable stars:

A dozen stars in this constellation are visible in small scopes, but most of them are of the EA type of eclipsing binaries, which change very little. These include two stars of the Trapezium (theta 1A and 1B).

> EA variables are old stars, nearing the end of their evolutionary process. The companion has grown to the size of a subgiant, perhaps equal in size to its primary. But their luminosities are quite different; thus, as the dimmer companion revolves around its primary, variations in the total brightness occur.
>
> The maximum brightness occurs of course when the two are not eclipsed, with each one adding its luminosity to the total output. Two minima also occur: the principal minimum is when the companion blocks out the primary; while a secondary minimum occurs when the companion is eclipsed by the primary.

The only interesting Mira-type regular variable is *U Orionis*, which usually has a brightness of 4.8 but every 368.3 days it drops down to 13. In 2000 the minimum is scheduled to occur on 5 December.

Deep Sky Objects:

M42, The Orion Nebula is perhaps the most photographed deep sky object in the heavens, a vast nebula of gas and dust exquisitely lit by surrounding stars.

This is a celestial nursery; soon (that's to say, in several hundred million years) young stars will appear from this wealth of cosmic matter.

Inside the nebula is the fascinating four-star system known as *The Trapezium*: theta 1A, 1B, 1C, and 1D - four stars held together by common gravity (actually at least two other stars are part of this complex system.) They are visible in medium sized telescopes and, with the nebula, form one of the most beautiful binary systems in the heavens.

M43 (NGC 1982) is a detached part of the Orion Nebula, with a ninth magnitude central star. A dark lane of gas separates M43 from M42, although the two are actually part of the same vast cloud.

M78 (NGC 2068) is a faint reflection nebula NE of Alnitak (zeta Ori), that looks best in long-exposure photographs.

The Horsehead Nebula is an intriguing and devilishly difficult dark nebula found just between zeta Orionis and sigma Orionis, visible in medium to large telescopes given the right sky conditions. An H-Beta filter is also helpful.

NOTES

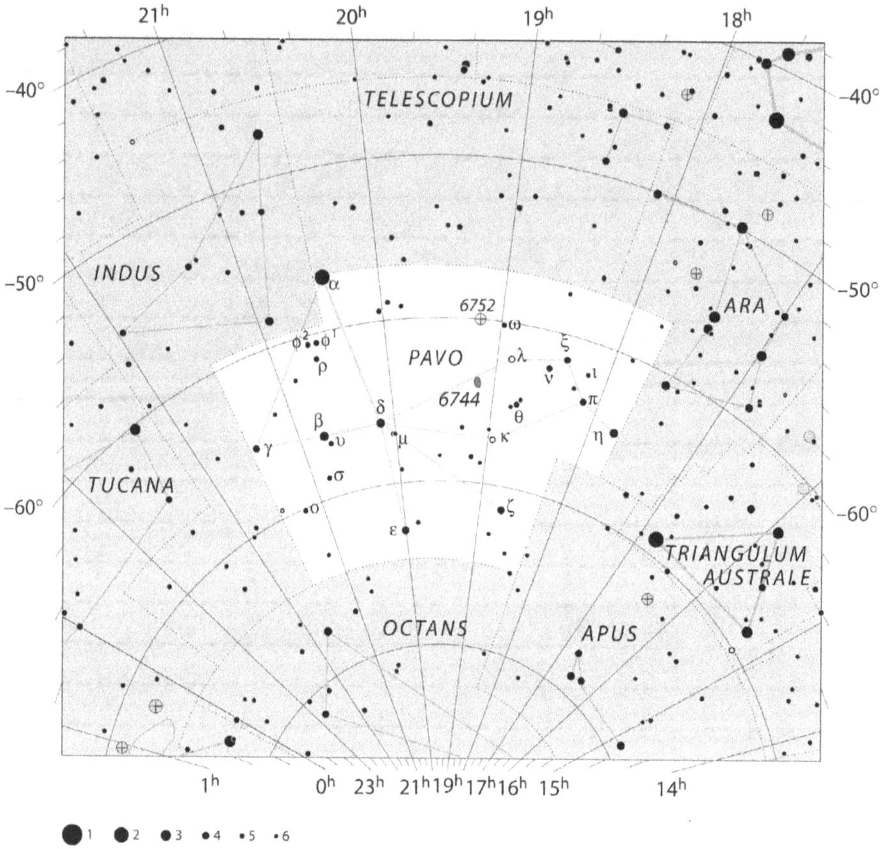

The Peacock

Hera, wife of Zeus and hence the Queen of the heavens, was an excessively jealous wife. And with good reason; Zeus was excessively amorous. Scholars have assiduously traced at least fifty lovers and mistresses of this supreme Greek god. Io was one of these lovers.

The trouble was, Io was one of Hera's priestesses, and Hera soon discovered the infidelity. To protect Io, Zeus transformed her into a heifer. But Hera was not fooled, and she claimed ownership over the heifer, then chose Argus Panoptes to guard the animal.

As indicated by its name, Argus Panoptes was "all eyes". Indeed, the beast had one hundred eyes, which surely should have been sufficient to guard one small heifer.

Zeus engaged Hermes with the task of rescuing Io. To avoid detection by one of Argus' one hundred eyes, Hermes charmed the animal with a flute when it was fast asleep, then threw a huge boulder on top of it, and for good measure cut off its head.

An angry Hera set a gadfly to pester Io, who then roamed around most of the Mediterranean nations before finally settling down in Egypt, where Zeus changed her back into human form. She later established the worship of Isis in Egypt.

As for the unfortunate Argus Panoptes, Hera put all of its many eyes on the tail of her sacred bird, the peacock. Only much later, in the seventeenth century, would the peacock itself become part of the heavenly zoo. Johann Bayer introduced the constellation in Uranometria in 1603, along with a number of other birds: Apus, Grus, Phoenix, and Tucana.

Pavo is a large constellation showing the tail of the peacock in full display. While the Bayer stars are not very bright, there are several deep sky objects of interest in the constellation.

Double stars:

Xi Pavonis is a visual binary: 4.4. 8; PA 154 degrees, separation 3.5".

Variable stars:

Kappa Pavonis is a cepheid: 3.91 to 4.8 every nine days.

Lambda Pavonis is a gamma Cas type variable: 4.0 to 4.3.

Deep Sky Objects:

NGC 6744 is a very large and fairly bright barred spiral galaxy. The galaxy is found three degrees SE of lambda Pavonis.

NGC 6752 is a splendid globular cluster, large and bright and compact. It's about ten degrees WSW of alpha Pav (omega Pav is just west to the west). It might be easier to find it four degrees north of NGC 7644. This cluster is considered one of the closest globulars, at about 20,000 light years away.

NOTES

Pegasus

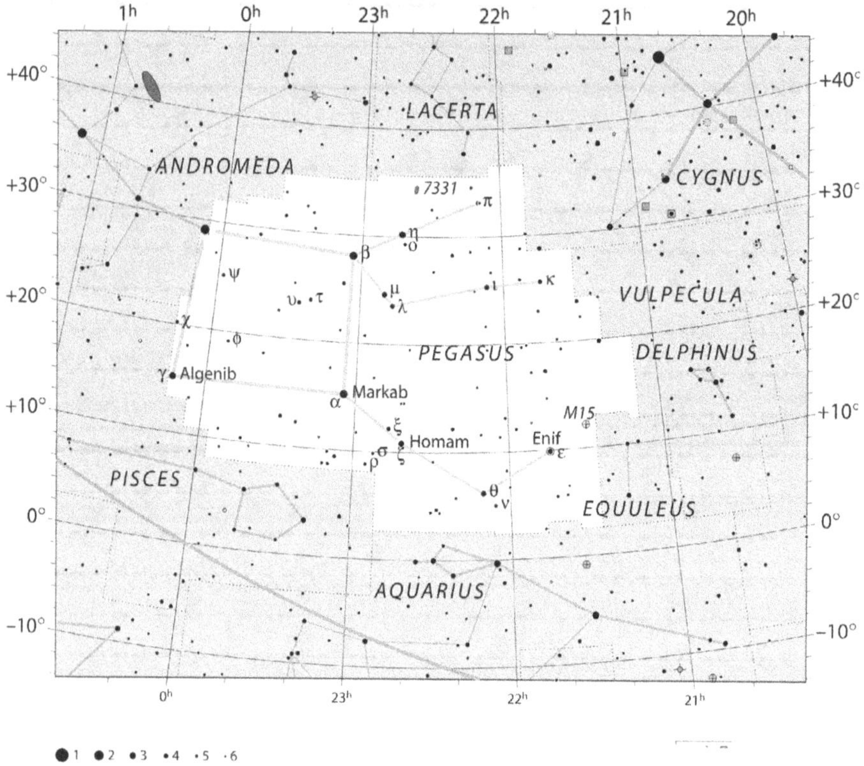

The Winged Horse

Pegasus, the winged horse, flew out of the head of Medusa when Perseus slew her. It was fathered by Poseidon, some time earlier, and waited for the Gorgon's death to appear. (Medusa's story is told under the constellation "Cepheus".)

Athene gave Pegasus to Bellerophon (a grandson of Sisyphus), who used the winged creature in his fight against the Chimaera - a monstrous female with three heads.

Pegasus is a conspicuous constellation which includes the so-called "Great Square of Pegasus". However it must now share the northeast corner of the square with Andromeda: delta Pegasus was given to Andromeda, to provide the lady with a head!

Pegasus went alone to Olympus, where he was used by Zeus to carry around his thunderbolts. As for Bellerophon, for his presumption of greatness, he wandered about the earth for the rest of his life, blind, lame, and shunned by man, until dying of old age. Pegasus is a conspicuous constellation which includes the so-called "Great Square of Pegasus". However it must now share the northeast corner of the square with Andromeda: *delta Pegasus* was given to Andromeda, to provide the lady with a head!

The stars are generally second and third magnitude. There are several interesting binaries here, a curious flare star, and one outstanding deep sky object.

Double stars:

Kappa Pegasi is a very close binary, with an orbit of

only 11.52 years: 4.8, 5.3; presently the companion is at PA 132 degrees and separation of only 0.2".

37 Pegasi is another close binary, with an orbit of 140 years: 5.8, 7.1; presently the companion is found at PA 118 degrees and separation of 0.8".

85 Pegasi is a well-known close binary with orbit of 26.27 years: 5.8, 8.9; currently the companion is at PA 149 degrees and separation of 0.8".

Variable stars:

Epsilon Pegasi is an irregular (Lb type) variable, and a flare star with a relatively cool shell. This supergiant can get as bright as 0.7 magnitude, and dimmer than 3.5. Generally it stays around 2.4.

Deep Sky Objects:

Pegasus has many galaxies and an outstanding globular cluster.

M15 (NGC 7078) is one of the finest globular clusters in the heavens, very bright and compact, at 35,000 to 40,000 light years away. It is found four degrees NW of epsilon Pegasi.

NGC 7331 is a spiral galaxy resembling the Milky Way Galaxy; it's as if we were looking at outselves from fifty million light years away.

NGC 7479 is a barred spiral galaxy about three degrees due south of alpha Pegasi.

NOTES

Perseus

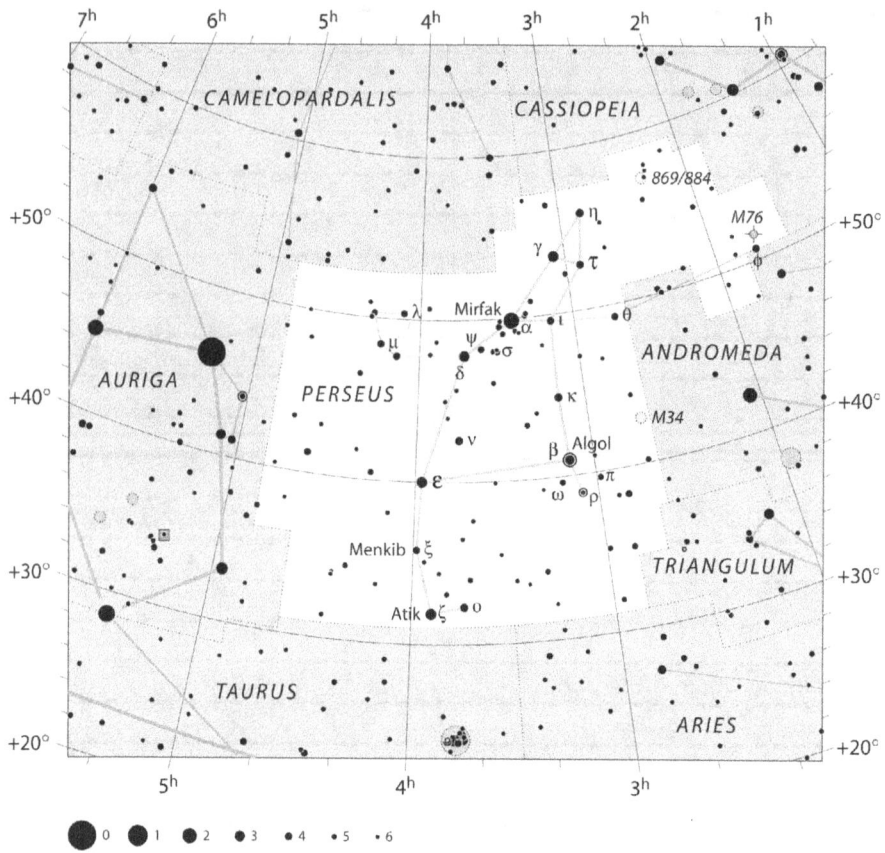

Perseus was one of the great heroes of classical mythology. He was the son of Zeus and Danae, and is best known for his killing of the Gorgon Medusa. This was a rather complex task, as anyone who saw her hideous face would be turned immediately to stone--the Gorgons, according to Bulfinch, were "monstrous females with huge teeth like those of swine, brazen claws, and snaky hair" (Bulfinch's Mythology, 109). Perseus accomplishes it, however, by the aid of Pluto, Mercury and Minerva. Pluto lent his helmet of invisibility to Perseus, Mercury lent the hero his winged sandals, and Minerva allowed him the use of her shield. With the aid of the helmet and the sandals, Perseus was able to get within striking range without being detected by Medusa or the two immortal Gorgons. He then used the reflection on the shield to guide his killing blow, and flew off unharmed bearing the head of Medusa:

He was bringing back the Gorgon's head, the memorable trophy he had won in his contest with that snaky-haired monster. As the victorious hero hovered over Libya's desert sands, drops of blood fell from the head. The earth caught them as they fell, and changed them into snakes of different kinds. So it came about that that land is full of deadly serpents. Thereafter, Perseus was driven by warring winds all over the vast expanse of sky: like a raincloud, he was blown this way and that. He flew over the whole earth, looking down from the heights of heaven to the land which lay far below (Metamorphoses IV 615-624).

He was rather tired and wanted to rest when he arrived at the lands of Atlas, at the ends of the earth. Atlas, however, tried to turn him away with his considerably greater strength. Perseus was infuriated and showed him the head of Medusa, turning the Titan into "a mountain as huge as the giant he had been. His beard and hair were turned into trees, his hands and shoulders were mountain ridges, and what had been his head was now the mountain top. His bones became rock. Then, expanding in all directions, he increased to a tremendous size--such was the will of the gods--and the whole sky with its many stars rested upon him" (Metamorphoses IV 656-662). Perseus flew on until he spotted the beautiful maiden Andromeda, who was chained to the rocky shore as a sacrifice to a sea monster. Perseus promptly fell in love with her, killed the monster, and married the princess.

When he died many years later, Perseus was immortalized as a constellation. He may be found near Andromeda and her parents, Cepheus and Cassiopeia, in the northern sky. The hero is depicted with a sword in one hand and the head of Medusa in the other; it is interesting to note the the eye of Medusa is the star Algol. Algol, which means "Demon Star" in Arabic, is an eclipsing binary star--it is normally about as bright as Polaris (second magnitude), but every two and a half days it becomes dimmer for roughly eight hours as the dimmer star of the pair passes between the brighter and the earth.

Perseus isn't a very clear asterism; some forms of the constellation have a straight line from alpha to theta, perhaps indicating Perseus' sword or sickle that he used to kill the Medusa.

However, as far back as Ptolemy's time, Perseus was seen as holding the head of Medusa, with Algol (beta Persei) being the "evil eye" of the Gorgon.

Perseus' stars are fairly bright. A good many of them go under other than Bayer names; several are notable binaries. There are also a few fine deep sky objects.

Double stars:

Epsilon Persei is rather difficult because of the dim companion: 2.9, 8.1; PA 10°, separation 8.8".

Zeta Persei is a multiple system, also with faint companions:

AB: 2.9, 9.5; 208° and separation 12.9".
C is a dim 11.3, PA 286° and separation 32.8";
D is 9.5, 195°, 94.2".

Eta Persei is a fixed triple system; AB are yellow and blue.

AB: 3.8, 8.5; PA 300°, separation 28.3".
C: 9.8; 268°, separation 66.6".

Struve 331 is a splendid fixed binary: 5.4, 6.8; PA 85°, separation 12.1".

It's found midway on a line between gamma Persei and tau Persei and just a bit south.

h1123 is a fine wide binary in the middle of M34: 8.5, 8.5; PA 248°, separation 20.0".

Variable stars:

Perseus had many types of variable stars, many of which are

too small to notice. Below are some of the more obvious examples.

Beta Persei (Algol) is a notable EA type eclipsing binary, changing from 2.12 to 3.39 every 2.8673 days (2d, 20h 48m 56s) as the companion eclipses the primary. The eclipse lasts roughly ten hours.

> The name itself, *Al Ghul*, means "Mischief Maker"; it is sometimes called Ras Algol, *Ra's al Ghul*, "The Demon's Head".
>
> This is a very bright white star, and the first eclipsing binary every discovered, in 1669 (thus giving the name "Algol variable" to this type of star). However the theory itself of an eclipsing binary being responsible for the variations in visual magnitude had to wait until 1783; this theory was only proved correct in 1889.
>
> The whole system is a bit more complicated than one star occulting another. A third component, Algol C, orbits both A and B about every 1.86 years, and even more companions have been proposed, but not proven.

R Persei is a long-period variable, 8.1 to 14.8 every 209.89 days. The next maximum is scheduled for mid May 1997 (then that December). In 2000 the maximum should occur the last week of October.

S Persei is an SRc type variable, 7.9 to 12 every 822 days (2.25 years). The next maximum should occur in the last half of October, 1998.

Deep Sky Objects:

M34 (NGC 1039) is a fine open cluster containing about eighty stars. The cluster is considered about 100 million years old. The cluster is about five degrees WNW of Algol (beta Persei), or more precisely twenty-seven arc minutes west of Algol and two degrees north.

NGC 869 and *NGC 884* form the well known "Double Cluster", two open star clusters side by side, easily seen by naked eye or binoculars.

The clusters are both considered babies, 869 only being about 6.5 million years old, and 884 about 11-12 million years old.

The easiest way to find them is to form a triangle, using gamma Andromedae and alpha Persei. Then the northern point becomes the twin clusters.

NGC 1499, The California Nebula, is a gaseous nebula one degree north of zeta Persei, and stretching itself in an east-west direction. Unfortunately it is extremely faint and difficult to view. In fact binoculars might afford the best chance.

The Perseid Meteor Showers

This meteor shower is active every year from late July to late August, primarily in the Northern Hemisphere. Peak showers occur on either 12 August or 13 August with about a meteor every minute at its peak. These showers derive from the Swift-Tuttle comet, which has an 130-year orbit.

Phoenix

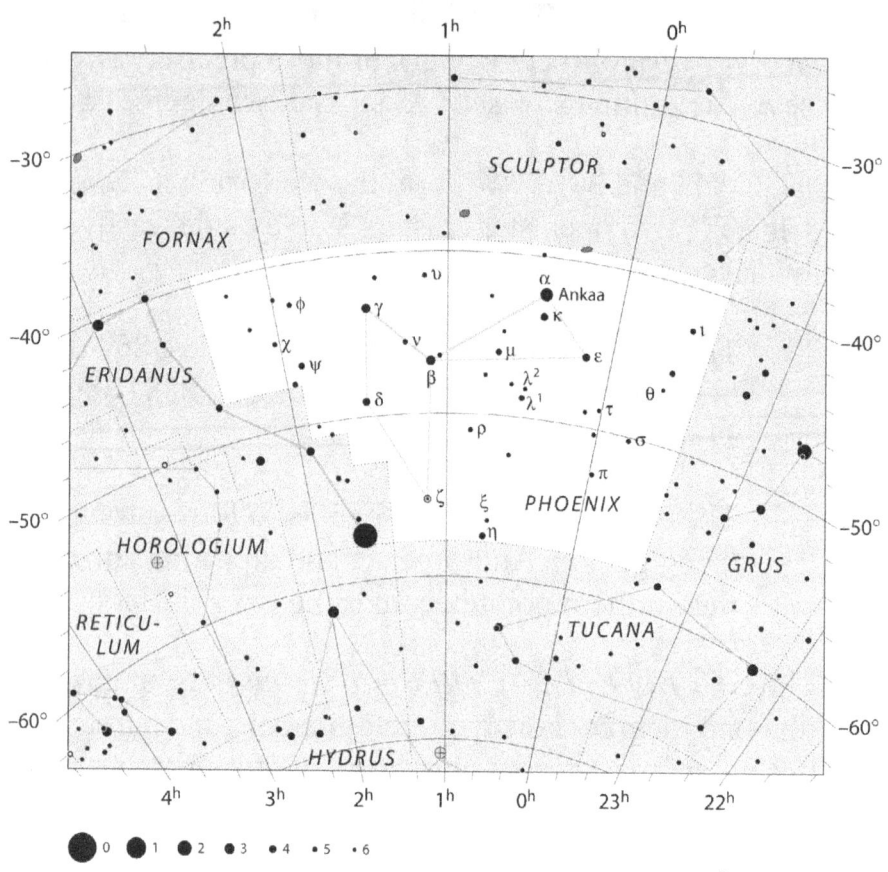

Fire Bird

Phoenix, the mythical bird rising from its own ashes, is another of those Southern Hemisphere constellations introduced by Johann Bayer in 1603.

In the past the constellation had been known as "The Boat" by the Arabs, then it became an eagle or other type of bird, so Bayer's decision to call it a phoenix does have some vague precedence.

The asterism is in fact rather like a large bird, rising into the air. The Bayer stars generally range from third to fifth magnitude.

There are a number of unspectacular binaries and several variables, including an important dwarf Cepheid.

Double stars:

Beta Phe is a multiple system:

AB: 4.2, 4.2; 346 degrees, 1.4".
C: 11.5 visual magnitude, at approx. 60 degrees and separation 57".

Zeta Phe is another multiple system:

AB: 4, 7; 40 degrees and separation 0.8".
C: 8; 243 degrees, 6.4".

Theta Phe: 6.5, 7; 275 degrees, 4.0".

Xi Phe has a faint companion: 5.7, 10; 253 degrees, 13.2".

Variable stars:

Rho Phoenicis is a delta Scuti type variable: 5.2-5.3 every 0.12 days (2h 38m).

SX Phoenicis is a dwarf cepheid with a considerable history.

> The star has a rather large proper motion (0.892"), but it is as a variable that the star is now better known.
>
> This is a pulsating A-type subdwarf with a period of slightly under eighty minutes (exactly 1h, 19m, 8.9s). The magnitude ranges from 6.7 to 7.5.
>
> The star is 2.5 degrees NE of *iota Phoenicis*.

Deep Sky Objects:

Phoenix has a number of galaxies, however they are all quite faint. The best of the group, NGC 55, is actually in nearby Sculptor.

NOTES

Pictor

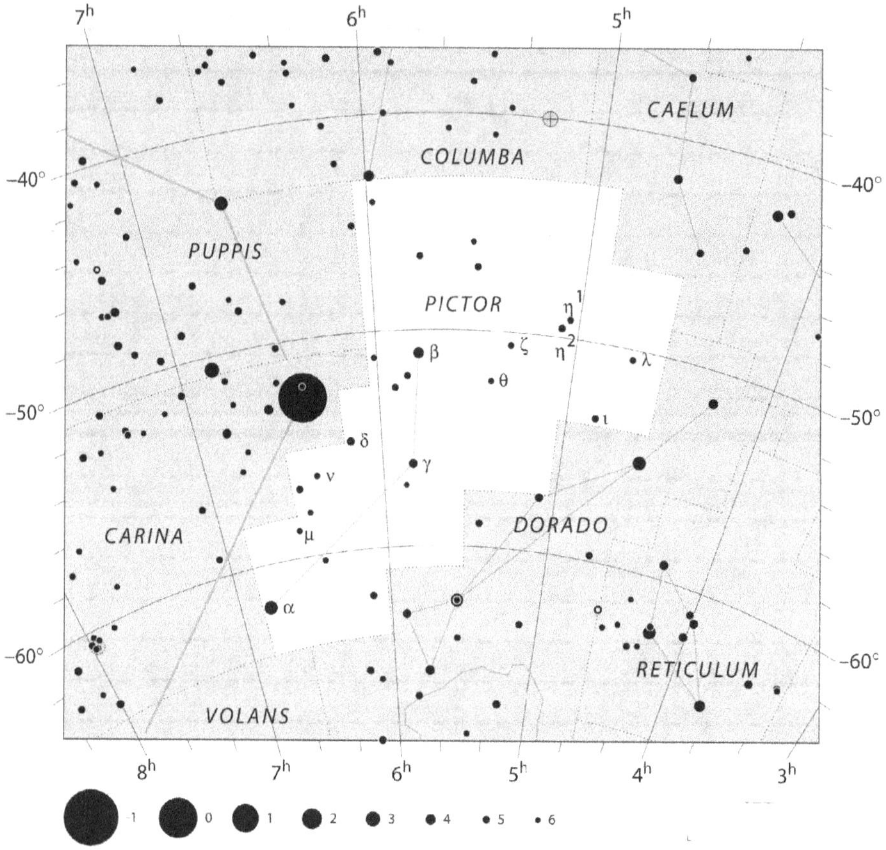

The Easel

One of the constellations representing technical and artistic apparatus that the Frenchman Nicolas Louis de Lacaille introduced into the southern sky after his observing expedition to the Cape of Good Hope in 1751–52. Lacaille's original title for the constellation, as given on his planisphere of 1756, was le Chevalet et la Palette, the easel and palette. In 1763 he Latinized this to Equuleus Pictorius (sic), while Bode in 1801 termed it Pluteum Pictoris. In 1844 the English astronomer John Herschel proposed shortening the name to Pictor, a suggestion adopted by Francis Baily in his British Association Catalogue of 1845. It has been known as Pictor ever since.

Pictor's stars are quite faint. Nevertheless there are several objects of interest, including the second fastest star known.

Double stars:

Theta Pictoris is a multiple system:

AB: 7.0, 7.5; PA 152°, separation 0.2"
C: 7.0; PA 287°, separation 38".

Iota Pictoris is a wide binary: 5.5, 6.5; PA 58°, separation 12.3".

Mu Pictoris has a relatively faint component: 6.0, 9.0; PA 231°, separation 2.4".

Variable stars:

Delta Pictoris is an eclipsing binary: 4.7 to 4.9 every 1.67 days.

R Pictoris is a semi-regular variable with a range from 6.9 to 10 about every 172 days.

Deep Sky Objects:

Kapteyn's Star is a fairly faint (8.8) red dwarf known for its large proper motion, which is second only to Barnard's Star (in Ophiuchus).

The star's discovery was made nearly a century ago, in 1897, by Professor Jacobus Cornelius Kapteyn (1851-1922) of the University of Groningen in Holland.

Kapteyn's lasting discovery was that all stars which have a measurable proper motion are part of one of two streams which move in different directions at different speeds.

Kapteyn's Star has a radial velocity of 242 km/s and a distance of 12.73 light years (=parallax of 0.256" or 3.91 pc). The star has a luminosity of only 0.004 suns (i.e. an absolute magnitude of 10.85). It is found roughly 8.5 degrees NW of beta Pictoris (Burnham, p. 1463, has a finder's chart).

The sun passes through the southeast corner of Pisces; in fact the vernal equinox now lies in Pisces.

Pisces is depicted as two fish connected by their tails at the star *alpha Piscium*. Indeed, alpha's name, "Al Rischa", means "the cord".

The constellation is rather faint; Pisces' stars are generally fourth magnitude. There are a few fine binaries, an interesting variable, and one Messier object: a splendid face-on spiral, which unfortunately is quite faint and rather a challenge for smaller telescopes.

Double stars:

Alpha Piscium (Struve 202) has an orbit of 933 years (considerably more than the 720 years previously thought): 4.3, 5.2; currently PA 223 degrees, separation 1.6".

Zeta Piscium (Struve 100) is a fine binary: 5.6, 6.5; 63 degrees, 23" separation.

Eta Piscium is a difficult binary to resolve: 3.5, 11; 36 degrees, 1" separation.

Psi^1 Piscium (Struve 88): 5.3, 5.5; 160 degrees, 30" separation.

Struve 61 (65 Piscium) is a splendid binary of equal stars: 6.3, 6.3; 297 degrees, 4.4" separation.

> The binary is found just on the border with Andromeda. The easiest way to find it is to start from zeta Andromedae, then move north 3 degrees and east half a degree.

Variable stars:

Kappa Psc is an alpha CV variable: 4.87-4.95.

TX Psc (19 Psc) is an interesting irregular, a deep red star that changes only slightly (about 5.0 to 5.5, although some references say from 5.5 to 6.0). Its main attraction is in the exceptionally deep redness of the star.

> The star is found between iota and lambda, north two degrees from lambda and one degree east. Or you might find it easier by first starting at gamma Psc and moving seven degrees east. (Burnham, p. 1475, has a

finder's chart.)

Deep Sky Objects:

The best deep sky object in Pisces is M74, the only Messier in the constellation.

> *M74 (NGC 628)* is a spiral galaxy seen face on. It's about 22 million light years away, and one of the faintest Messiers. The larger the scope, the better. Long exposure photographs show two or three loosely-wound spirals `spinning' out from a small bright nucleus.
>
> The galaxy is found 1.5 degrees ENE of eta Piscium.

NOTES

Pisces

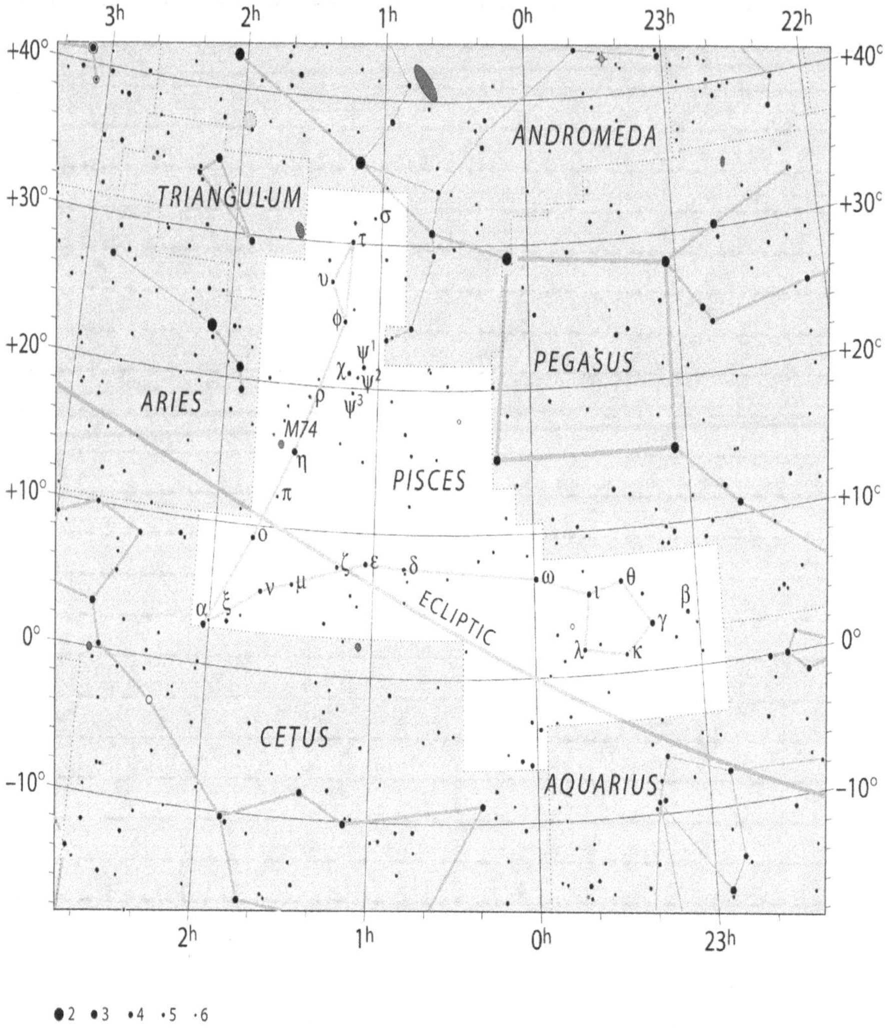

The Two Fish

The horrible earthborn giant Typhoeus suddenly appeared one day, startling all the gods into taking on different forms to flee. Zues, for instance, transformed himself into a ram; Hermes became an ibis; Apollo took on the shape of a crow; Artemis hid herself as a cat; and Bacchus disguised himself as a goat. Aphrodite and her son Eros were bathing on the banks of the Euphrates River that day, and took on the shapes of a pair of fish to escape danger. Minerva later immortalized the event by placing the figures of two fish amongst the stars.

The zodiacal constellation Pisces represents two fish, tied together with a cord. The constellation is neither particularly bright nor easy to locate, but it lies near Pegasus and Aquarius.

The sun passes through the southeast corner of Pisces; in fact the vernal equinox now lies in Pisces.

Pisces is depicted as two fish connected by their tails at the star *alpha Piscium*. Indeed, alpha's name, "Al Rischa", means "the cord".

The constellation is rather faint; Pisces' stars are generally fourth magnitude. There are a few fine binaries, an interesting variable, and one Messier object: a splendid face-on spiral, which unfortunately is quite faint and rather a challenge for smaller telescopes.

Double stars:

Alpha Piscium (Struve 202) has an orbit of 933 years (considerably more than the 720 years previously thought): 4.3, 5.2; currently PA 223 degrees, separation 1.6".

Zeta Piscium (Struve 100) is a fine binary: 5.6, 6.5; 63 degrees, 23" separation.

Eta Piscium is a difficult binary to resolve: 3.5, 11; 36 degrees, 1" separation.

Psi^1 Piscium (Struve 88): 5.3, 5.5; 160 degrees, 30" separation.

Struve 61 (65 Piscium) is a splendid binary of equal stars: 6.3, 6.3; 297 degrees, 4.4" separation.

> The binary is found just on the border with Andromeda. The easiest way to find it is to start from zeta Andromedae, then move north 3 degrees and east half a degree.

Variable stars:

Kappa Psc is an alpha CV variable: 4.87-4.95.

TX Psc (19 Psc) is an interesting irregular, a deep red star that changes only slightly (about 5.0 to 5.5, although some references say from 5.5 to 6.0). Its main attraction is in the exceptionally deep redness of the star.

> The star is found between iota and lambda, north two degrees from lambda and one degree east. Or you might find it easier by first starting at gamma Psc and moving seven degrees east. (Burnham, p. 1475, has a finder's chart.)

Deep Sky Objects:

The best deep sky object in Pisces is M74, the only Messier in the constellation.

M74 (NGC 628) is a spiral galaxy seen face on. It's about 22 million light years away, and one of the faintest Messiers. The larger the scope, the better. Long exposure photographs show two or three loosely-wound spirals `spinning' out from a small bright nucleus.

The galaxy is found 1.5 degrees ENE of eta Piscium.

NOTES

Piscis Austrinus

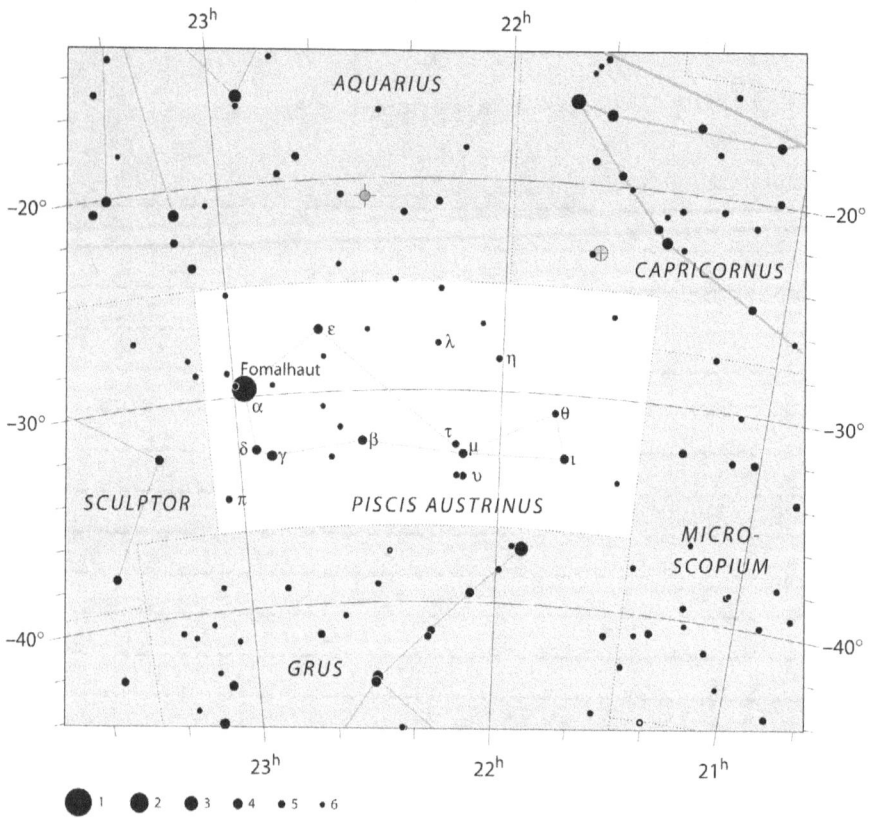

The Large Southern Fish

Piscis Austrinus, also known as Piscis Australis, is a fish lying on its back, drinking in the waters pouring from the jars of Aquarius. The asterism leaves a lot to the imagination.

The constellation was known in ancient times, and is said to be the original "Pisces". It is thought that the constellation referred to the Assyrian fish god Dagon and the Babylonian god Oannes. Even the Arabs called the constellation Al Hut al Janubiyy (The Large Southern Fish).

The bright star Fomalhaut (alpha PsA) was so named from Fum al Hut, meaning "The Fish's Mouth", although it carried many other names as well, including Al Difdi al Awwal ("The First Frog").

> The bright star Fomalhaut (alpha PsA) was so named from *Fum al Hut*, meaning "The Fish's Mouth", although it carried many other names as well, including *Al Difdi al Awwal* ("The First Frog").

There are several fine binaries here, as well as a red dwarf with the fourth highest proper motion. Except for Fomalhaut, all other stars in this constellation are fourth or fifth magnitude.

Double stars:

Piscis Austrinus has several visual binaries, either fixed or moving very slowly.

Beta PsA is a fixed binary: 4.5, 7.5; 172 degrees, 30.4".

Delta PsA is a fairly wide but faint binary: 4.5, 10; 244 degrees, 5".

Gamma PsA: 4.5, 8.5; 262 degrees, 4.3".

Eta PsA: 5.5, 6.5; 116 degrees, 1.6".

Variable stars:

Piscis Austrinus has no notable variables.

Deep Sky Objects:

NGC 7172 is a spiral galaxy seen almost edge-on; there is a dark equatorial band seen with larger scopes. The galaxy is two degrees NW of mu PsA.

In the same field, just to the south, lie three more galaxies: 7173, 7174, and 7176.

Lacaille 9352 is a red dwarf 11.68 light years away, with a visual magnitude of 7.44. It has a proper motion of 6.901", fourth highest (after Barnard's Star, Kapteyn's Star, and Groombridge 1830).

Lacaille 9352 is found one degree SE of pi PsA; Burnham has a finder's chart.

Double stars:

9 Puppis is a very close binary with rapid orbit of only 23.18 years. Currently the values are: 5.6, 6.2; PA 315°, separation 0.4".

k^1 Puppis and *k^2 Puppis* form a noted system of nearly equal stars: 4.5, 4.7; PA 318°, separation 9.9".

Note that the name here is "k" not "kappa"; many of Puppis' stars are English labels.

NOTES

Puppis

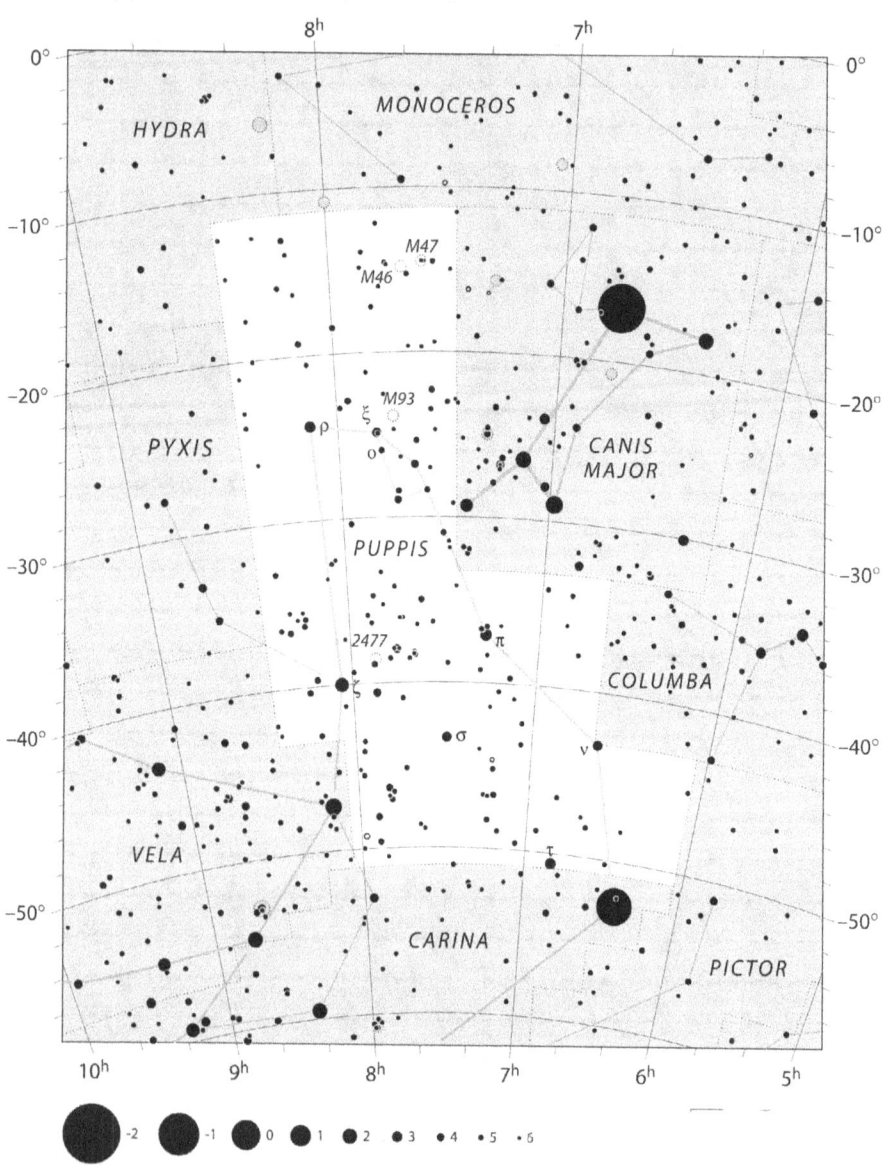

The Stern

Puppis, "The Stern", is the largest constellation associated with the former constellation "Argo Navis", the Argonauts' Ship.

It was Nicolas Louis de Lacaille who dismantled the older constellation in the mid-eighteenth century, breaking it into four smaller constellations: Carina, Pyxis, Puppis, and Vela.

Not only quite large, the constellation spans a rich area of the Milky Way, guaranteeing the amateur astronomer a number of fine objects to study.

Double stars:

9 Puppis is a very close binary with rapid orbit of only 23.18 years. Currently the values are: 5.6, 6.2; PA 315°, separation 0.4".

k^1 Puppis and *k^2 Puppis* form a noted system of nearly equal stars: 4.5, 4.7; PA 318°, separation 9.9".

Note that the name here is "k" not "kappa"; many of Puppis' stars are English labels.

Variable stars:

Rho Puppis is a delta Scuti type variable: 2.68 to 3.87 every 3h22m52s.

L_2 Puppis is a noted semi-regular variable with a wide range, from 2.6 to 6.2 about every 141 days. This is a giant reg star 150 light years away. Note that the label carries a *sub* rather than the normal super..., although some references do not follow the majority in this

matter.

Its unrelated neighbour, L_1, is an alpha CV variable: 4.86 to 4.93, every 22h.

Deep Sky Objects:

Puppis has three Messiers and several more deep sky objects of interest.

M46 (NGC 2437) is a fine open cluster of perhaps five hundred stars about 4000-5000 light years away. Sitting on the northern edge of the cluster is a planetary nebula, NGC 2438, which is about 3000 light years away.

> The cluster is found in the northern portion of the constellation, eleven degrees east of Sirius (alpha CMa) and two degrees north.

M47 (NGC 2422) is a bright open cluster in the same field as M46, just one degree west of M46. Of the two, M47 is the brighter, as it includes several fifth and sixth magnitude stars.

M93 (NGC 2447) is another open cluster, quite bright but smaller than the two previous objects. It's found 1.5 degrees NW of xi Puppis

NGC 2477 is a very fine globular cluster three degrees NW of zeta Puppis, nearly half way between pi Puppis and zeta Puppis.

NOTES

Pyxis

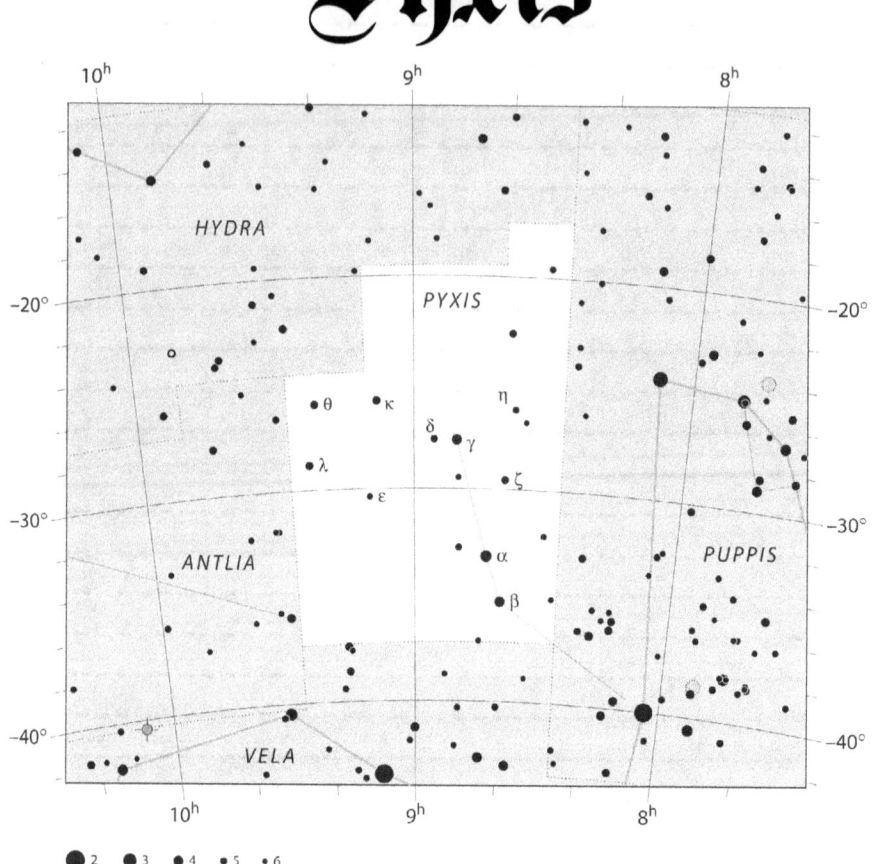

The Compass

Pyxis, "The Compass", is one of Nicolas Louis de Lacaille's creations. He called it "Pyxis Nautica" (The Nautical Box, or Mariner's Compass).

The constellation's Bayer stars range from third to fifth magnitude.

Double stars:

Zeta Pyxidis is a wide binary with faint companion: 4.9, 10; PA 61°, separation 52.3".

Epsilon Pyx is a fairly wide binary also with a rather faint companion: 5.6, 9.5; PA 147°, separation 17.8".

Kappa Pyx: 4.6, 10; PA 262°, separation 2.1".

NOTES

Reticulum

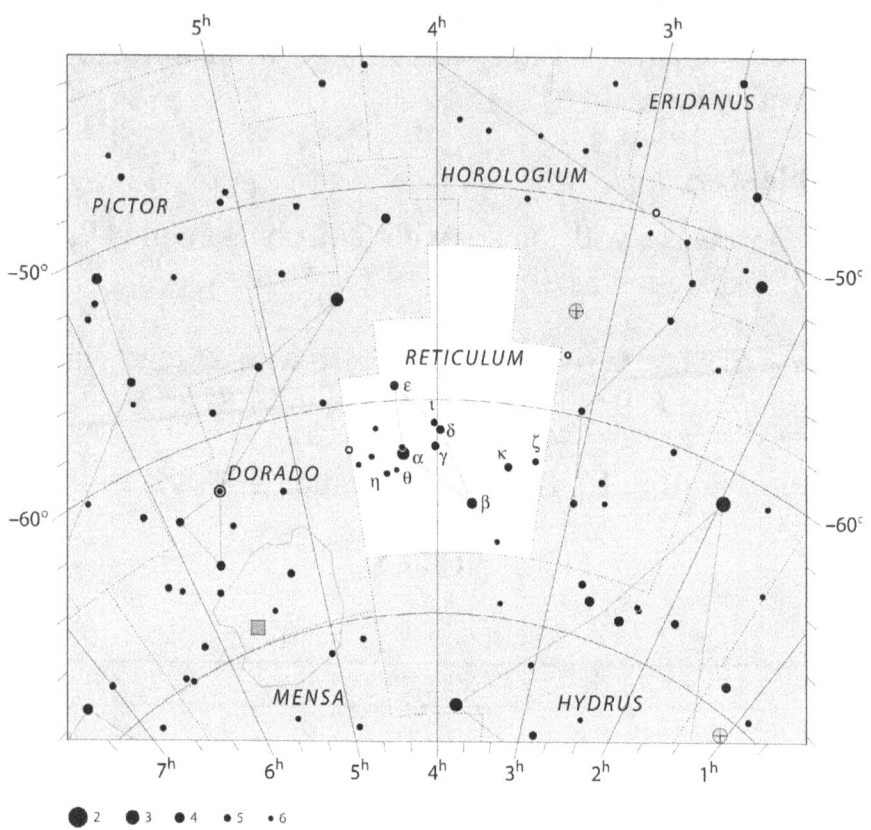

The Reticule

Reticulum was introduced by Nicolas Louis de Lacaille in the mid 1700s, meant to commemorate the reticule, an instrument used by Lacaille to measure star positions.

Reticulum is a very small and rather bleak constellation, with stars in the third to fifth magnitude range.

Double stars:

Zeta Reticuli is a wide visual binary of two yellow stars: 5.2, 5.5; PA 222 degrees, separation 130".

Theta Reticuli: 5.9, 8.0; PA 4 degrees, separation 4.1".

Variable stars:

Gamma Reticuli is a semi-regular: 4.42-4.64 every 25 days or so.

Deep Sky Objects:

NGC 1559 is a large and fairly bright spiral galaxy seen face on. It's a Seyfert galaxy (i.e. it's a source of significant nonthermal emissions, such as x-rays and ultaviolet). The galaxy is found between alpha and theta Ret.

NOTES

Sagitta

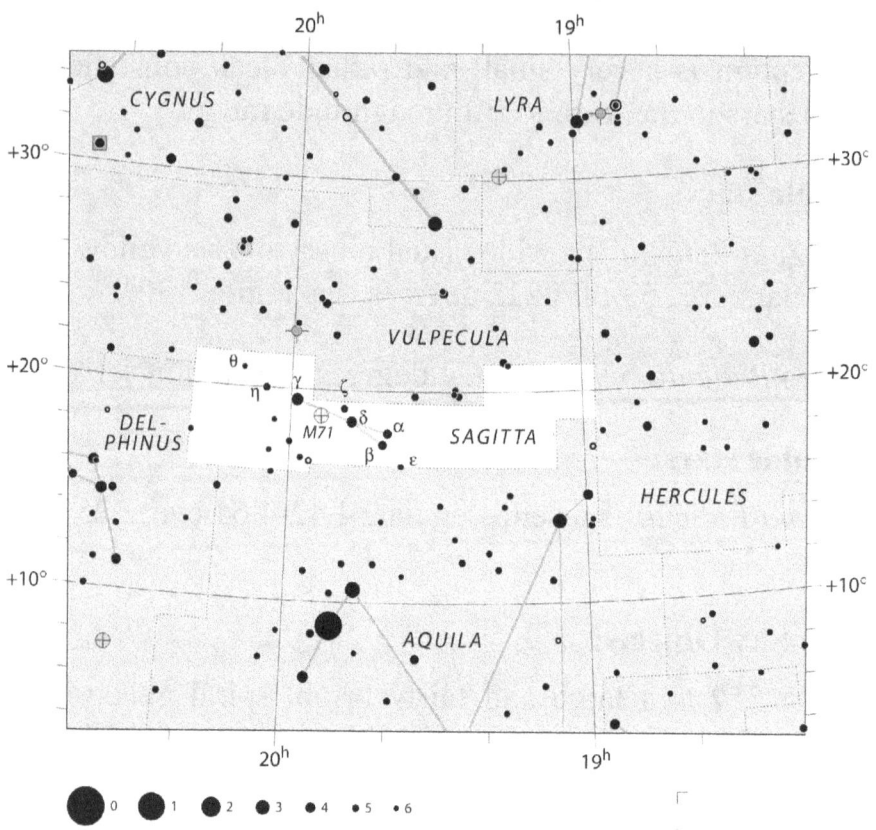

The Arrow

𝔖agitta, "The Arrow", while small and insignificant, is a constellation known to the Greeks. Some references believe that Sagittarius, the Archer, shot the arrow (apparently without a known target); others talk about Cupid, and Heracles, and Apollo. The point is, there isn't any established myth associated with Sagitta.

The constellation has a number of interesting items, including several multiple binary systems and a Messier object.

The Bayer stars range from 3.5 to 6.5. Note that the brightest star here is gamma Sagittae.

Double stars:

Zeta Sge (Struve 2585) is a close binary with 22.78 year orbit; it's also a multiple system:

> **AB (1962 values): 5.5, 6.5; PA 180°, separation 0.2"**
> **C: 9; PA 311°, separation 8.4"**
> **D: 11; PA 247°, separation 75".**

Theta Sge (Struve 2637) is also a multiple system:

> **AB: 6.5, 8.5; PA 325°, separation 12"**
> **C: 7, PA 223°, separation 84".**

Variable stars:

S Sagittae is a fairly bright cepheid, ranging from 5.5 to 6.2 every 8.38 days.

To locate S Sge, drop three degrees south of gamma Sge. The bright star here is 11 Sge. S Sge is in the same field, just to the west-southwest.

U Sagittae is a well-known Algol-type eclipsing variable, suitable or small telescope or binoculars.

Use the nearby star just to the NE (*Struve 2504*, visual magnitude 7.9) as a comparison; U Sge is slightly brighter than this star at its regular brightness, then dips far below (to about 9.2) every 3d9h8m5s as the larger companion completely eclipses the primary. This lasts for about 1h40m, then the star rapidly regains its brightest magnitude.

U Sagittae is five degrees west of alpha Sge, and 1.75 degrees to the north. It forms an equilateral triangle with two brighter stars, 1 Vulpeculae and 4 Vulpeculae. (Note: on Tirion's *Sky Atlas*, U Sge appears to have the label "OU" due to the size of the star itself.)

Deep Sky Objects:

M71 (NGC 6838) is classified as a globular cluster, but it looks much more like an open cluster. It's found just between delta and gamma Sagittae and slightly south.

NOTES

Sagittarius

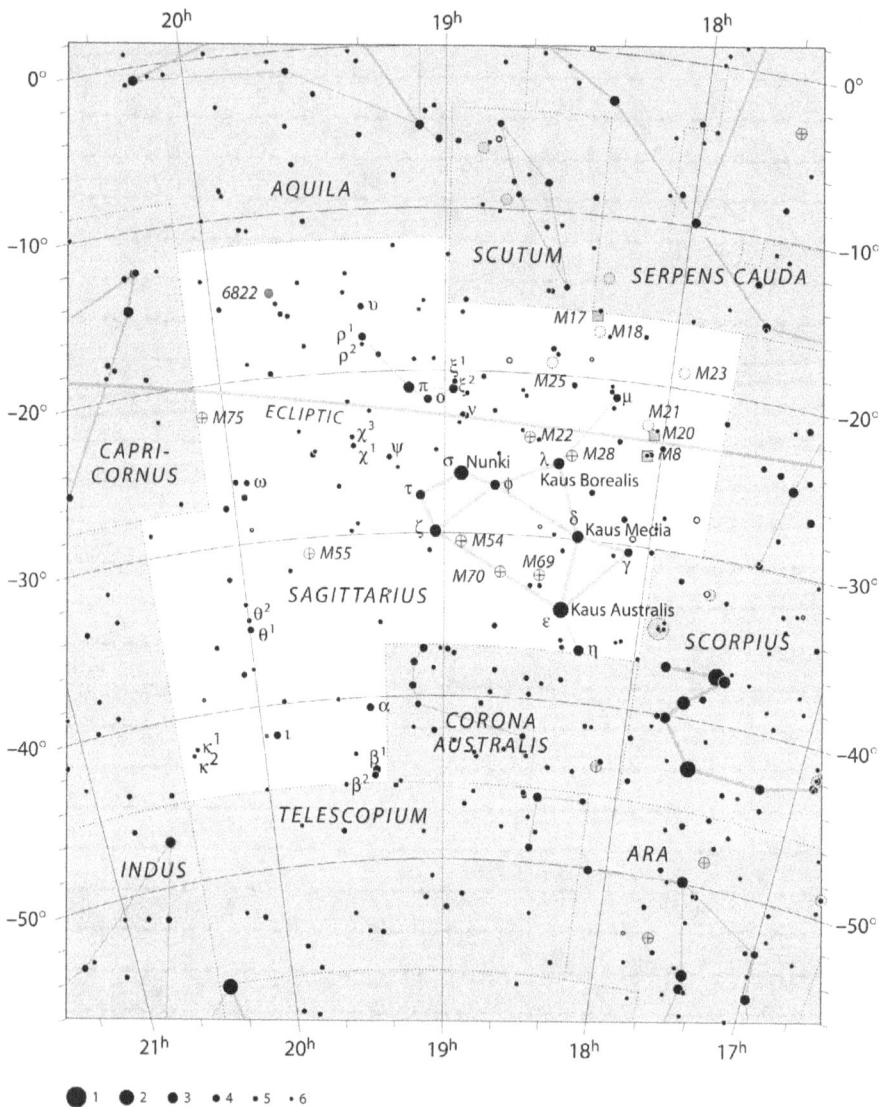

The Archer

The zodiacal constellation Sagittarius represents the centaur Chiron. Most of the centaurs were regarded in myth as bestial--they were, after all, half horse. However, the ancient Greeks had a great deal of respect for the horse, and so were reluctant to make the centaurs entirely bad. In fact, Chiron was renowned for his gentleness. He was an excellent archer, musician, and physician, and tutored the likes of Achilles, Jason, and Heracles.

Chiron, however, was accidentally shot and wounded by Heracles. The arrow, which had been dipped in the poison of the Lernaean Hydra, inflicted great suffering on Chiron--so great, in fact, that even the talented physician could not cure himself. In agony, but as an immortal unable to find release in death, Chiron instead offered himself as a substitute for Prometheus. The gods had punished Prometheus for giving fire to man by chaining him to a rock. Each day an eagle would devour his liver, and each night it would grow back.

It was the Romans who named the constellation Sagittarius ("sagitta" is Latin for `arrow'), although several stars carry Arabic names which identify just which portion of the constellation they represent:

> *Alpha Sagittarii* is named "Rukbat": (Rukbat al Rami=Archer's knee), and *beta Sgr* is "Arkab" (Tendon).

> The bow is outlined by three stars:

>> *Lambda Sgr*: "Kaus Borealis" = the northern (part of the) bow

>> *Delta Sgr*: "Kaus Meridionalis" = the middle (part

of the) bow

Epsilon Sgr: "Kaus Australis" = the southern (part of the) bow

The arrow tip is *gamma Sgr* ("Al Nasl" = the point)

While the asterism of the bow is quite apparent, it takes some imagination to see the half-man, half-beast pulling back on the string. Perhaps it helps to know that *zeta Sagittarii* is named "Ascella" (the armpit of the archer), while *nu Sgr* is "Ain al Rami": The Eye of the Archer.

The Bayer stars are generally third and fourth magnitude. The brightest star is *epsilon Sgr*, while *alpha Sgr* is nearly fourth magnitude. In fact, there are fourteen stars brighter than *alpha*).

The constellation has a number of fine binaries, and several superb deep sky objects.

Double stars:

Nu^1 Sagittarii is a fixed binary with faint companion: 5.0, 10.8; PA 97° and separation 2.5".

Note that *nu^1* and *nu^2* are not gravitationally bound, although they form an optical binary of some historical importance: these two stars caused Ptolemy to write about "a nebulous double star" long before Hershel coined the term "binary".

54 Sgr also catalogued as *h 599* is a multiple system:

AB: 5.4, 12; PA 274°, separation 38"; AC: 8.9; PA 42°, 45.6". The primary has a reddish tinge to it.

Rho¹ and *rho²* form a nice triangle with *h 2866*:

AB: 8.0, 8.3; 53°, 23.4"
AC: 8.6; 137°, 24".

Variable stars:

Sagittarius has a variety of variables, some of which are suitable for small scopes, primarily cepheids but also one Mira-type long range variable.

Upsilon Sgr is an eclipsing binary (beta Lyrae type: EB) with an unusually long period of 137.9 days. Its range will be undetectable to most observers, from 4.53 to 4.61, but what makes the system interesting is that it seems to be one of the most luminous systems known (with an estimated absolute magnitude of around -7.5).

The brightest cepheids are: *W Sgr* (4.3-5.1 every 7.6 days) and *X Sgr* (4.2-4.9 every 7 days).

R Sagittarii is a long-period variable fluctuating from 6.7 to 12.8 every 269.84 days. In 2000 the maximum should occur in the second week of July.

The star is found two degrees NE of *pi Sagittarii*, or just past the midpoint of a line between *pi* and *rho Sgr*.

Deep Sky Objects:

Sagittarius has fifteen Messier objects, far more than any other constellation. However these fifteen are of varying quality. Three are spectacular, and a number of others are bright and impressive but a number are quite disappointing. While they are all included here, due to space limitations the less interesting objects have been omitted from the constellation graphic.

M8 (NGC 6523) is a marvellous diffuse nebula known as the "Lagoon Nebula".

> This naked eye object is considered to be from 3500 to 5100 light years away. A dark band divides the nebula in two. While easily spotted with the eye, there is a wealth of detail that can only be brought out with at least a medium sized scope.
>
> The open cluster NGC 6530 is contained in the eastern part of the nebula. The young cluster (only several million years old) is nicely contrasted against the nebula.
>
> The Lagoon Nebula is five degrees west of *lambda Sgr* and one degree north.

M17 (NGC 6618), the "Swan Nebula" or the "Omega Nebula", and occasionally known as the "Horseshoe Nebula". This nebula resembles the tail of a comet: a bright diffuse trail of light with a bit of a hook on it. It is about 5000 light years away.

> The Swan Nebula is five degrees north of *mu Sgr*, and one degree east.

M18 (NGC 6613) is an open cluster of about twenty

stars; a rather undistinguised member of the Messier group found one degree south of M 17.

M20 (NGC 6514), the "Trifid Nebula", is another delight, but only with larger scopes, which will bring out the three dark lanes familiar on photographs. In the same field is M 21, an open cluster of about fifty stars.

> The Trifid Nebula is found 1.5 degrees north of the Lagoon Nebula.

M21 (NGC 6531) is a rather unspectacular open cluster 0.7 degrees NW of M20.

M22 (NGC 6656) is a fine globular cluster, a highly concentrated group of perhaps five hundred thousand stars in total, about 20,000 light years away. It lies two degrees NE of *lambda Sgr*.

M23 (NGC 6494) is a pleasantly scattered open cluster of about 120 stars located four degrees northwest of *mu Sgr* and one degree north.

M24 (no NGC) is a bright "star cloud", which contains the open cluster NGC 6603.

M25 (no NGC) is a bright open cluster but without much interest.

M28 (NGC 6626) is a bright condensed globular cluster, much less spectacular than M 22 but a fine object none the less. It is one degree NW of *lambda Sgr*.

M54 (NGC 6715) is a globular cluster, difficult to

resolve.

M55 (NGC 6809) is another globular cluster, less concentrated than those previously mentioned. It is about 20,000 light years away, and lies between *zeta Sgr* and *theta Sgr*: seven degrees east of *zeta* and one degree south.

M69 (NGC 6637) is a globular cluster of little merit.

M70 (NGC 6637) is another globular cluster, two degrees east of M69. It too is of little interest.

M75 (NGC 6637) is the faintest of globular clusters in this constellation.

NGC 6822, "Barnard's Galaxy". Very faint; the larger the scope the better. This irregular dwarf galaxy is about 1.7 million light years away, making it one of the closest of its kind. It's in the same region as 54 Sgr, six degrees northeast of *rho Sgr*.

> Since Sagittarius sits at the very heart of the Milky Way, there are many more deep sky objects to study: planetary nebulae abound, as well as both bright and dark nebulae and of course star clusters, especially of the globular variety.

NOTES

Scorpius

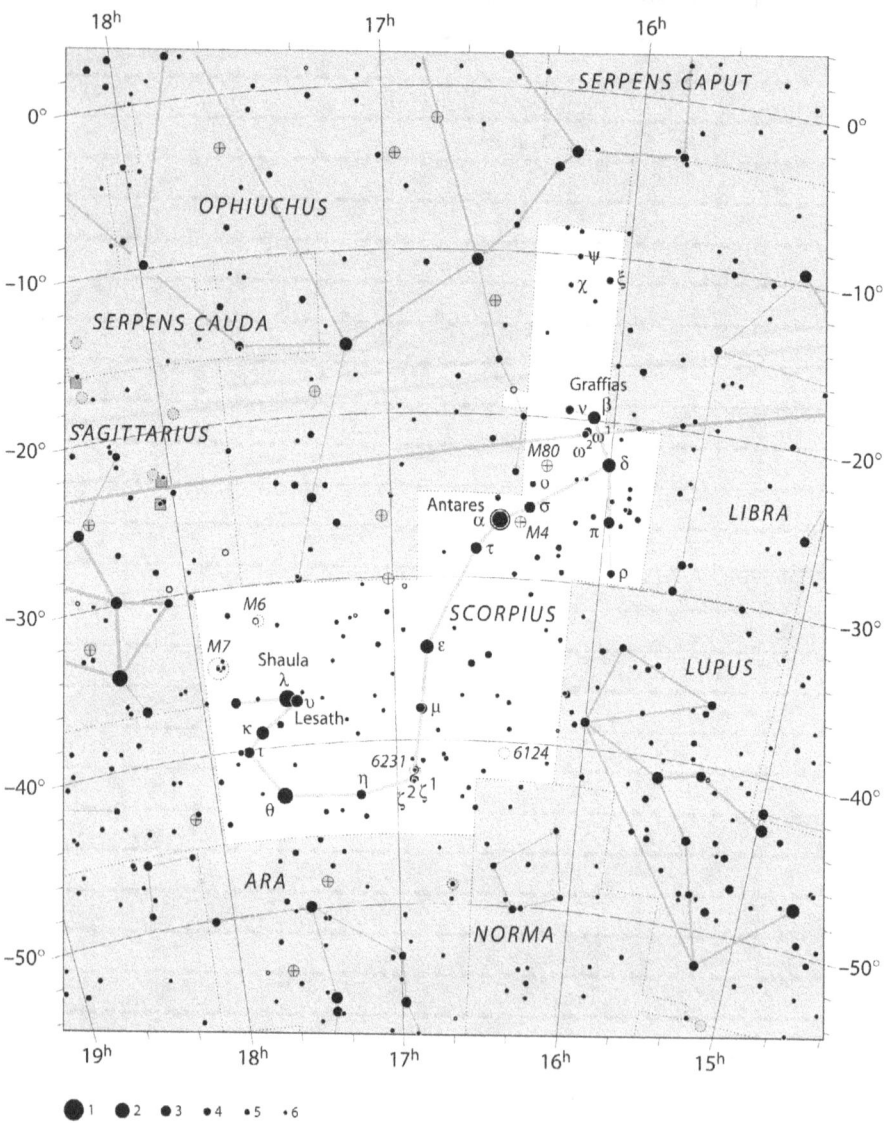

The Scorpion

Scorpius is a zodiacal constellation. The scorpion is generally believed to be responsible for the death of the great hunter Orion. According to some myths, the scorpion stung Orion in response to his boast that he could defeat any beast; according to others, it was sent by Apollo, who was concerned for his sister Artemis's continued chastity.

In either case, Scorpius was placed in the opposite side of the sky from Orion so as to avoid any further conflict. It is to the southeast of Libra, and is marked by the bright red star Antares. (Antares is Greek for "Rival of Ares," the Greek war-god. The star is so named because of of its brightness and color, which are approximately the same as of the planet Mars.

Scorpius is one of the oldest constellations known - possibly even one of the original six signs of the zodiac. While the sun still traverses Scorpius, it only takes nine days to do so; most of the time is spent in neighbouring Ophiuchus (which is the only constellation that the sun enters but which is not a part of the zodiac).

The asterism of a gigantic skewed "S" was seen in many ancient cultures as a scorpion, possibly handed down by cultural conquest or influence. The two stars lambda and upsilon, both called "The Sting" in Arabic, traditionally form the stinger, although some star maps currently show the nearby "G Scorpii" as one of the stingers. We have recently changed our graphic to reflect the original stingers.

The constellation was once much larger, but the western portion representing the claws of the scorpion was given to Libra.

The star table indicates just how bright many of Scorpius's stars are; in fact the constellation is one of the brightest of the larger constellations.

Alpha Scorpii is better known as *Antares* ("Rival of Mars"). This is one of the four Royal Stars of the ancients, along with Aldebaran, Regulus, and Fomalhaut. It glitters with an unusual metallic red while the entire region is bathed in a pale red nebula, lit from the same star.

> This red supergiant has a visual binary that just might be visible, depending on local conditions and the size of one's scope (see below). The star is estimated to be between 285 sun diameters to about 700 suns. It's 600 light years away.

> Due west 1° (about half the distance to sigma Sco) is the bright globular cluster M4, while another globular cluster, M80, is 4° NNW of Antares. See below for these deep sky objects.

Double stars:

Alpha Scorpii is a visual binary which may be difficult to resolve due to the brightness of the primary. Try a moonlight night, which should cut the glare of the brighter star: 1.1, 5.4; PA 274°, separation 2.6".

> The companion is usually described as green in colour, probably a visual effect created by the red glow of Antares. The star is estimated to orbit its primary every 900 years.

Beta Scorpii. This superb double has a pleasant colour contrast: white and bluish-green. 2.6, 4.9; PA 23°, 13.7".

Nu Scorpii is a multiple system, a "double-double". That is, each of the visible components (AC) is also a primary of a closer component; these are termed AB and CD.

> **AC: 4.4, 6.4; 337°, and 41" separation.**
> **AB: 4.4, 5.4; PA 2°, 1.3".**
> **CD: 6.3, 8.0; 51°, 2.3".**

Xi Scorpii is also a multiple system, a system which also includes the next binary system as well (Struve 1999).

> Components AB form a close binary with period of 45.7 years. The companion is now gradually drawing away from the primary: PA 308° and separation 0.39".

Sigma Scorpii: a double with faint companion. AB: 2.9, 8.5; PA 273°, separation 20".

Struve 1999 is gravitationally attached to the Xi Scorpii system, although at a distance of about 7000 AU (an "AU"-- astronomical unit-- being the distance of the earth from the sun).

The binary is found just south of Xi Scorpii, two yellow stars of nearly equal brightness: 7.4, 8.1; 99°, 11.6".

Variable stars:

RR Scorpii is the brightest long-period variable in the constellation, with a visual magnitude range of 5.0-12.4 every 281.45 days. In 1999 the maximum should occur around the end of May.

Deep Sky Objects:

There are four Messier objects in Scorpius (some authorities put a fifth in the constellation as well: M62, but usually it is listed in Ophiuchus).

M4 (NGC 6121) is a rather near globular cluster (6000-10,000 light years) but without a large telescope it will not appear very spectacular. There may be as many as fifty RR Lyrae variables in the cluster.

> M4 is located just west of Antares, roughly half way to sigma Scorpii.

M6 (NGC 6405) is the second-best cluster of the constellation (after M7). This is an open cluster which sometimes bears the name "The Butterfly Cluster". Its brightest star is BM Scorpii, a sixth-magnitude yellow giant. The cluster is about 1500-2000 light years away.

M7 (NGC 6475) has no name, but is clearly the best deep sky object of the constellation. This magnificent open cluster is extremely large (two full-moon diameters) and quite bright, being visible even to the naked eye under the right conditions.

> A scope easily resolves the stars, the brightest twenty-two of which range from 5.6 to 9.0. There are several close visual binaries in the cluster. (See Burnham for these, as well as extensive notes on this cluster.)

> M7 is 4° NNE of lambda Scorpii. It's about 800 light years away.

M80 (NGC 6093) is a rather faint, very compact, globular cluster in the vicinity of Antares, between this star and beta Scorpii, and more narrowly speaking, nearly midpoint between two 8th-magnitude stars (which are the brightest stars of the region). The cluster is quite distant, some 36,000 light years away, and it takes a very large telescope to study it in detail.

NGC 6231 is a naked-eye open cluster one half degree north of zeta Scorpii (which is in fact a member of the group). This cluster is certainly worthy of being a Messier; while noticeable to the naked eye, binoculars resolve its various members. It's about 5500-6000 light years from us.

> The stars that make up the cluster are generally supergiants that resemble the Pleiades in miniature. Burnham points out that if this cluster were the same distance as the Pleiades, its stars would outshine the Pleiades "by a factor of about 50 times".
>
> The cluster is only part of a much larger, very scattered, cluster called H 12, which is found one degree north. In fact, the stars seen as joining NGC 6231 and H 12 actually form one of the spiral arms of our own galaxy.

NOTES

Sculptor

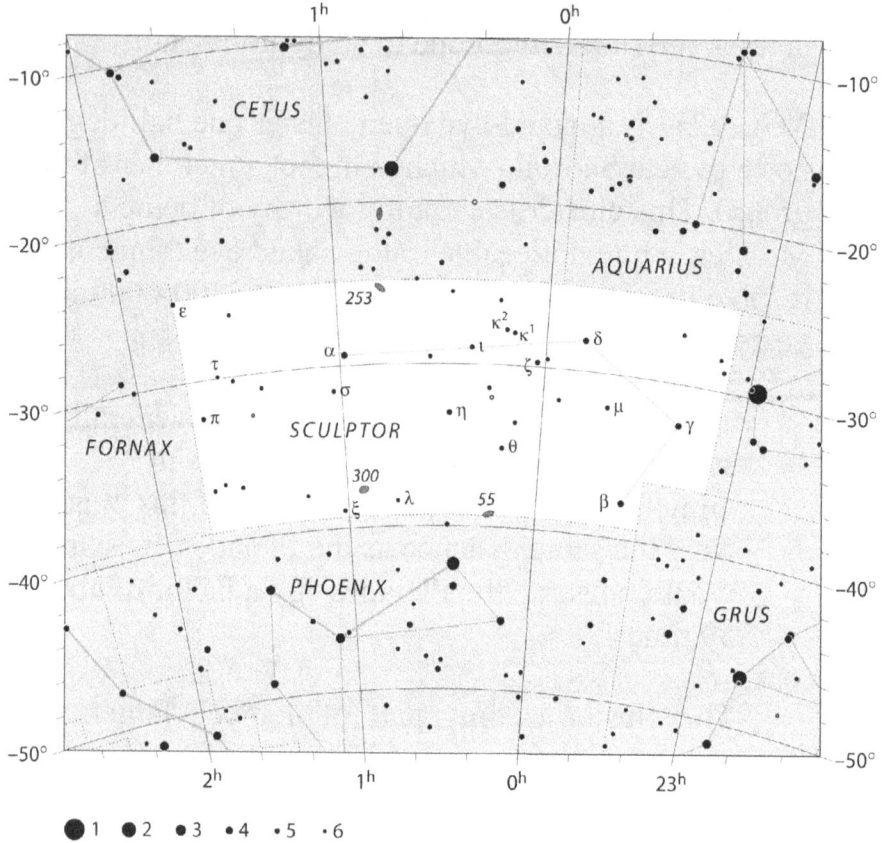

The Sculptor

Sculptor is one of those obscure constellations invented by Nicolas Louis de Lacaille to help fill in part of the southern sky. The full name was "L'Atelier du Sculpteur", the Sculptor's Studio.

Its stars are generally fourth and fifth magnitude. The constellation is has two binaries with very long orbits, a couple of nice variables, and several spiral galaxies.

Double stars:

Delta Sculptoris is a multiple system:

> **AB: 4.6, 11.5; 243°, 4".**
> **C: 9.5; 297°, 75".**

Epsilon Sculptoris is a slow moving binary with an orbit of about 1200 years: 5.4, 8.6; currently 23°, 4.7".

Kappa^1 Scl: 6.1, 6.2; 265°, 1.4".

Tau Scl is another slow moving binary; it takes nearly 1900 years to make one revolution: 6.0, 7.1; 340°, 2.1".

Variable stars:

R Sculptoris is a deep red semi-variable, ranging from 9.1 to 12.9; the period is roughly 370 days.

> The star lies almost midway between sigma and pi Scl, slightly toward pi Scl.

S Sculptoris is a bright Mira-type variable ranging from 5.5 to 13.6. It has a period of 362.6 days, which means that it brightens up at just about the same time every year. Thus, until the year 2000 it should reach its maximum during the last week in August; in 2000 the maximum should occur in mid-August.

The star is three degrees NNE of theta Sculptoris.

Deep Sky Objects:

NGC 55 is a spiral galaxy, seen nearly edge-on. It's located twelve degrees southwest of *alpha Scl*.

The nearest star to this galaxy is alpha Phoenicis, which lies three and a half degrees southeast.

This is a member of the so-called Sculptor Group, which is one of the nearest galaxy clusters closest to the Milky Way, at about 8 million light years.

NGC 253 is considered one of the easiest spiral galaxies to observe, apart from M31 in Andromeda. This is perhaps the brighest spiral of the Sculptor Group. It is found five degrees NNW of alpha Sculptoris.

NOTES

Scutum

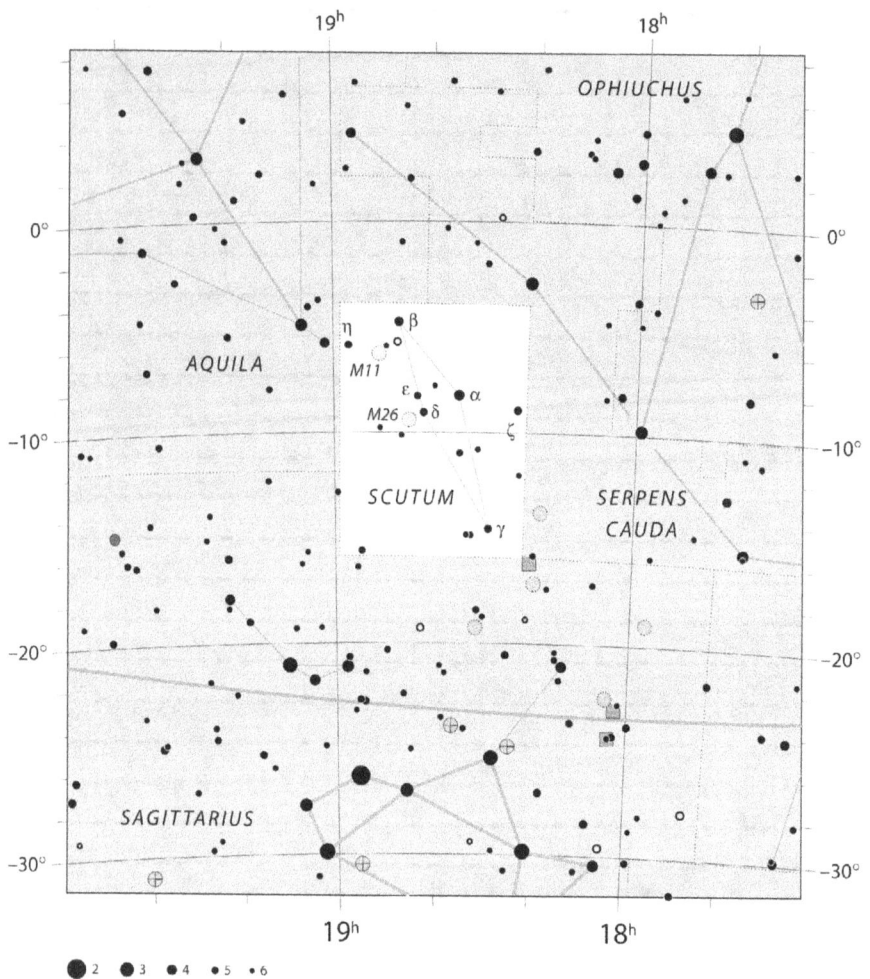

The Shield

Scutum was invented by the Polish astronomer Johannes Hevelius, and was placed in his posthumous catalog of 1690, the Prodromus Astronomiae, along with Canes Venatici, Lacerta, Leo Minor, Lynx, Sextans, and Vulpecula.

These newer constellations became better known after being accepted by John Flamsteed in his catalog published in 1725.

The proper name is Scutum Sobiescianum, Sobieski's
Shield, as the constellation pays honour to Jan Sobieski.

Jan Sobieski (1629-1696) was the eldest son of the castellan of Crakow, Jakob Sobieski. He was a brilliant military leader and by 1665 had become the field commander of the Polish army.

The main threat to Poland at this time (indeed to all of central Europe) came from the Turks. While Sobieski attempted to repulse the Turks, the Polish king's envoys ceded all the Ukraine to Turkey. Meanwhile Sobieski won victory after victory. In November of 1673 the king died. Sobieski left the front lines and presented himself as a candidate for the throne back in Warsaw. (The kingship was an elected position.) In May of 1674 he became King John (or Jan) III.

Sobieski returned to his former job as army commander, and after nearly a ten year struggle, he was able to sign the Treaty of Warsaw with Leopold I. Following this treaty, Sobieski further safeguarded Europe from the Turks. Personally leading the Polish cavalry, on 12 September 1683, he broke the Turkish siege on Vienna, and liberated Hungary in the bargain.

Seven years later Hevelius commemorated these events with the inclusion of Scutum Sobiescianum in the heavens.

The asterism, faint as it is, does resemble a simple shield, complete with handle. Some see a dipper as well, standing on its handle. For this you need to include delta and eta Sct.

Scutum is quite dim, with few Bayer stars. There are two Messier objects however and an important variable.

Double stars:

Delta Scuti has a rather faint companion: 4.5, 12; PA 46 degrees, separation 15". An optical component is sometimes also given as part of this system (C: 10; PA 130°, 52.5").

Epsilon Scuti is a multiple system of very faint companions.

> **AB: 5, 14.5; PA 97°, 13.6".**
> **C: 13.5; 195°, 37.6".**
> **D: 14.5; 312°, 15.4".**

Variable stars:

Delta Scuti is the prototype of a class of variables. Its range is 4.6-4.79 and it has a period of four hours thirty-nine minutes.

Delta Scuti variables are giant stars with spectra of A to F (occasionally G). They resemble dwarf Cepheids in that they are also pulsatig variables, but they have a smaller amplitude: the delta Scuti amplitude may typically be less than a tenth of a magnitude.

These stars are quite young and are frequently found in open clusters. The brightest delta Scuti variable is not delta Scuti, but rather beta Cassiopeiae.

R Scuti is the brightest RV Tau type variable: 4.2-8.6, period 146.5 days.

RV Tau stars form another class of pulsating variables, a kind of semi-regular Cepheid. There are not many known, perhaps a hundred or so (the other well known member is R Sagittae). They are supergiants with spectra of F to K (M in some instances) and they are all quite luminous. They are often associated with globular clusters.

Deep Sky Objects:

M11 (NGC 6705), "Wild Duck Cluster", is a fine open cluster of perhaps four hundred stars which fan out like a flight of startled mallard.

The cluster is one degree SE of R Scuti. (Another rather nice binary, Struve 2391, is found between R Scuti and M11: 2.6, 9; PA 333, 38".)

M26 (NGC 6694) is another open cluster in Scutum; about thirty stars that resemble a miniature horseshoe.

The cluster is one degree SE of delta Scuti.

Serpens Caput

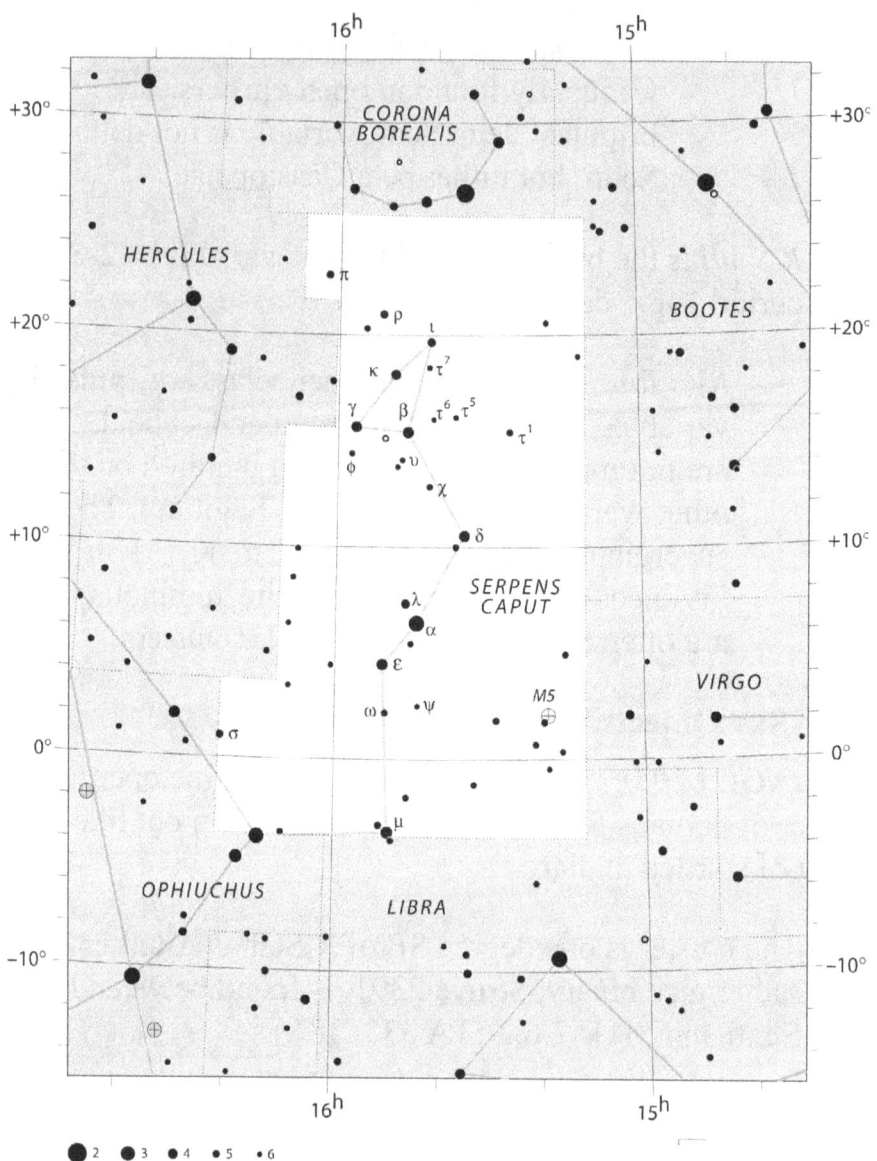

The Serpent

This is the second part of the Ophiuchus- Serpens group. The Serpent is being grasped in the hands of Ophiuchus the Serpent Holder. Thus the constellation wraps around Ophiuchus, and is divided into two parts: Serpens Caput (the head) and Serpens Cauda (the tail).

The constellation Serpens is spread across a greater part of the sky than is Ophiuchus, but it has far fewer features of interest. Still, there are several Messier objects and some very nice binaries.

There isn't any remarkable asterism in Serpens, and it might take some effort to decide just which stars belong to this constellation, and which belong to Ophiuchus. That is, the Bayer stars of Serpens compare in brilliance with those of Ophiuchus.

The brightest star, *alpha Serpentis*, is called *Unukalhai*, meaning "Neck of the Snake". It is 67 light years away, and is approximately ten times the size of the sun.

Double stars:

Serpens has three visual binaries of some interest, two of which are very attractive, and one which will test your observing skills.

Beta Serpentis (Struve 1970) is a wide visual yet difficult to observe due to the brightness of the primary compared to the faint companion: 3.0, 9.2; PA 265°, separation 30.8".

Theta Serpentis (Struve 2417) is a wonderful binary of two white stars: 4.0, 4.2; 103°, 22.2".

Struve 2375 is a superb pair: 6.2, 6.6; 116°, 2.4".

Variable stars:

R Serpentis is a long-period variable, 5.2-14.4, with a period of 356.41 days. In the year 2000 the star should have had a maximum on about 5 February.

The star is located 1.2 degrees ESE of beta Serpentis, nearly midway between beta and gamma Serpentis and very slightly south of a line drawn between them.

Deep Sky Objects:

There are two Messier objects in Serpens: M5 and M16; the first is found in the "head" of the serpent, the second in the "tail".

M5 (NGC 5904) is a spectacular globular cluster, containing a half a million stars. The cluster is quite compact and rather bright; it is about 25,000 light years away, and ten billion years old.

The cluster is found eight degrees SW of alpha Serpentis.

M16 (NGC 6611), "The Eagle Nebula", is a remarkable open star cluster surrounded by a huge nebula, very luminous with dark streaks of dust: a nursery of newly forming stars. Best seen in large scopes; a nebula filter might help.

The cluster is fifteen degrees south of eta Serpentis, but an easier way to find it may be to draw a line from eta Serpentis to xi Serpentis, to the SW. Now midway along that line is found the bright star nu Ophiuchi. Draw an imaginary perpendicular out from nu Ophiuchi, southeast. About seven degrees along this line is M16.

If this seems a bit complicated, first try locating M17, The Omega Nebula (or Swan Nebula), in Sagittarius. Two and half degrees north is M16.

NOTES

Sextans

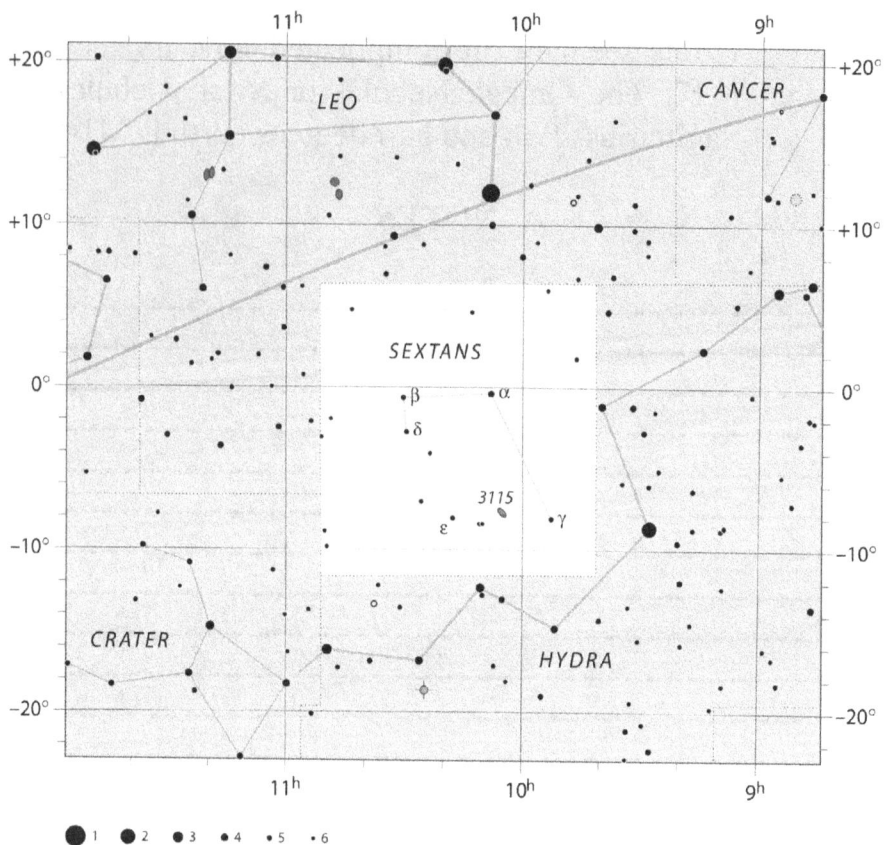

The Sextant

Sextans, "The Sextant" is one of a number of constellations devised by Johannes Hevelius, and published posthumously in 1690.

Rather than rely on the telescope, Hevelius usually used the sextant to do his viewing. Thus he commemorates the instrument here. The simple asterism shows two straight lines. The Bayer stars here are fragmentary and only fourth and fifth magnitude.

Double stars:

Gamma Sextantis is a multiple binary.

AB is a close double, with orbit of 77.5 years: 5.6, 6.1; 61°, 0.5".
C is much fainter and wider: visual magnitude 12; 325°, 36".

Variable stars:

There are no variable stars of interest to the amateur astronomer.

Deep Sky Objects:

NGC 3115 is a bright galaxy seen edge-on, looking like a fuzzy flying saucer. It may be over 20 million light years away.

The galaxy lies midway between epsilon Sextantis and gamma Sextantis and very slightly north. The easiest way to find it is start from Regulus (alpha Leonis). From this bright star drop south through alpha Sextantis (12.5 degrees) then continue south another 7.5 degrees.

Taurus

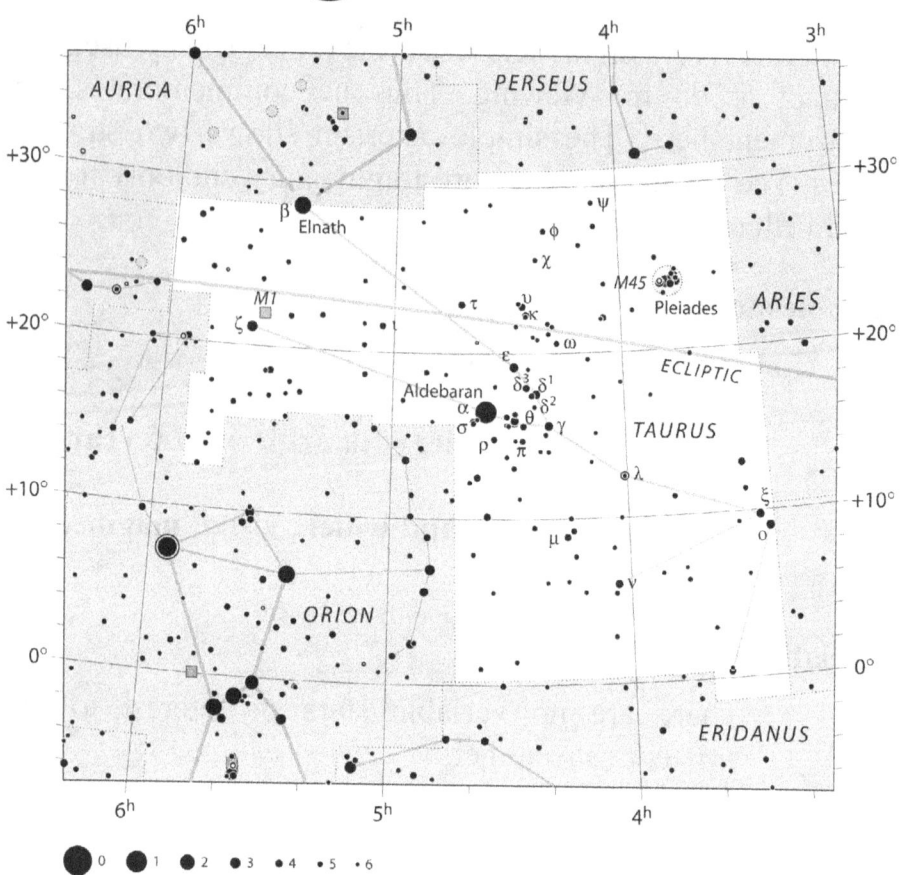

The Bull

𝔗aurus represents the bull-form taken on by Zeus when he became enamored of Europa, princess of Phoenicia:

> Majesty and love go ill together, nor can they long share one abode. Abandoning the dignity of his sceptre, the father and ruler of the gods, whose hand wields the flaming three-forked bolt, whose nod shakes the universe, adopted the guise of a bull; and mingling with the other bullocks, joined in their lowing and ambled in the tender grass, a fair sight to see. His hide was white as untrodden snow, snow not yet melted by the rainy South wind. The muscles stood out on his neck.

The constellation shows mainly the horns, and exceedingly long horns they are. The left (southern) horn starts from the group of stars known as The Hyades, of which Aldebaran seems (erroneously) to be a member. It extends from Aldebaran to zeta Tauri, near the eastern edge of the constellation.

The right horn lifts up just west of the Hyades, from delta Tauri through tau Tauri and finally to its tip at beta Tauri (El Nath: remember this star as part of Auriga?)

The rest of the bull is rather disappointing; a slight body and two spindly legs. It may be that the bull is half-emerged in water, as it carries Europa across to Crete.

The stars of Taurus:

Taurus' eye is bright and piercing. This is *Aldebaran (alpha Tauri)*, an orange giant about 40 times the size of the Sun. Aldebaran is an old star. For billions of years it has burned its supply of hydrogen until there is little left. Its future won't be as a spectacular explosion of a supernova but rather a gradual dimming into a white dwarf.

Following the lower horn out to its tip we find *zeta Tauri*. This is a shell star. Shell stars are main-sequence stars which rotate rapidly, causing a loss of matter to an ever-expanding shell.

Most of the interesting features of Taurus are found in the centre of the constellation and toward the west. Around Aldebaran are a number of stars which go by the collective name of *The Hyades* (see below).

Aldebaran is not a member of this group. Not only is it closer to us, but its proper motion is at a different angle. Aldebaran is moving at an angle of 161 degrees, the stars of the Hyades at around 102-109 degrees.

Double stars:

Taurus has an abundant selection of binary stars, including many Struve binaries that we haven't mentioned. Below is a very small selection of some of the easier doubles to resolve.

Theta^2 and *theta^1* form a fixed binary of wide separation, *theta^2* just below and to the east. Note that *theta^2* is the primary: 3.4, 3.8; PA 346° and separation 337".

Kappa^1 and *kappa^1* form an easily resolved binary: 4.2, 5.3; PA 328°, separation 5.3".

Sigma^2 and *sigma^1* is another wide fixed binary. And again, *sigma²* is the primary: 4.8, 5.2; PA 193° and separation 431".

80 Tauri is a difficult visual binary with an orbit of 189.5 years: 5.5, 8.0; current PA 17° and separation of 1.8" (very nearly its maximum separation).

Struve 422 is a wide visual binary with an orbit of over 2000 years: 5.9, 8.8; PA 269°, 6.7". It's located at 9° SW of *nu Tauri*, just north of the brighter *10 Tauri*.

Variable stars:

Many of the more notable variable stars in Taurus are of a type not noticed by casual observation, such as *alpha Taurus*, which is classified as an Lb type variable. These are irregular giants whose variation can only be detected by means of photoelectric photometry. *Alpha Taurus* only changes in visual magnitude by 0.2, from 0.75 to 0.95, and the period is irregular.

BU Tauri (Pleione) is a gammaCas type variable, from 4.77 to 5.50. GammaCas variables are also characterised by an irregular period, which may sometimes be very rapid. These are B stars, quite young, and rotate very rapidly. This rotation results in the throwing off of material, which then forms a shell around the star. The cause of its variation is still not understood.

Zeta Tauri is also a gammaCas type variable, with a variation from 2.88 down to 3.17 roughly every 133 days.

Lambda Tauri, in the Hyades cluster, is a good example of an eclipsing variable. The variability is caused by the partial eclipse of the primary by its companion, dimming the 3.3 visual magnitude down to 3.8 every 3.95 days.

R Tauri is a Mira-type variable with a 320.9 day period. Usually at 7.6, it drops to a very dim 15.8 once a year. In 2000 the maximum should occur in the first week of May.

Deep Sky Objects:

Taurus contains two well known Messier objects: the Crab Nebula and the Pleiades. Besides these two there is the `other' cluster, known as The Hyades, and the curious "Hind's Variable Nebula".

Just northwest of zeta Tauri is the first of Messier's objects: M1, the *Crab Nebula*. Early observers thought the object to be a star cluster, something like a dimmer version of the Great Orion Nebula. Messier was so intrigued by it, on the night of 12 September, 1758, that he began his catalogue - the purpose of which was to keep observers from mistaking such objects for comets.

It takes a rather large telescope to see any of the filamentary features of the nebula; most viewers come away disappointed.

The Crab Nebula is a remnant of a supernova, whose explosion occurred (or rather, was visibly recorded) in July of 1054. Chinese and Japanese astronomers witnessed the event. In fact, it would have been difficult not to notice, for the night sky would have been lit up by a star with the visual magnitude of about -5, bright enough to be seen even in the daytime for nearly a month.

The star that exploded, producing the nebula, is now an optical pulsar. Even now, nearly a thousand years later, the nebula is hurtling through space at roughly a thousand kilometers per second. And it continues to grow; the nebula is now over thirteen light years in diameter (four parsecs) according to the *Facts On File Dictionary of Astronomy*.

M45, The Pleiades:

This open cluster contains as many as three thousand stars. The brightest seven go under the name The Seven Sisters" (from brighter to dimmer): Alcyone (eta Tauri), Electra, Maia, Merope, Taygeta, Celaeno, and Asterope. Added to the list are also Pleione (BU Tauri = 28 Tauri), just east of Alcyone, and Atlas (27 Tauri) who are actually Mum and Dad for the seven sisters. (The two are often seen as one star; it takes a clear night to see them as two separate stars.)

The Hyades

This open cluster of about two hundred stars is only 150 light years away, and considered to be about 600 million years old. It is shaped like a "V", just to the west of Aldebaran.

Just as the Pleiades have individual names, so did the Hyades at one time. In fact, these stars were supposed to be the half-sisters of the Pleiades, and Robert Burnham (*Celestial Handbook*) gives their names - and a great deal more on this group. Theta^2 is the brightest star of the group, which forms a binary with theta^1 (see below). The group is thought to be about 400 million years old.

These nine stars, then, constitute the minimum count, easily seen with the naked eye, while there are actually as many as 250 stars which belong to the group. The cluster is estimated to be 415 light years away. Even a small telescope brings this famous star cluster alive.

Hind's Variable Nebula (NGC 1555)

This curious deep sky object is located two degrees west of epsilon Tauri, and two degrees north of delta Tauri. First look for the rather dim variable T Tauri. Burnham (*Celestial Handbook*) has a finder's chart, on page 1833. The star has an irregular variability, from 9 to 13.

Very close to T Tauri, just off to the west, is a cloud-like object. This is Hind's Variable Nebula. Its variability is long-lasting; from 1869 to 1890 it couldn't be found at all. Presently, it seems to be gaining slightly in visual magnitude, although its actual visual magnitude hasn't been determined.

NOTES

Telescopium

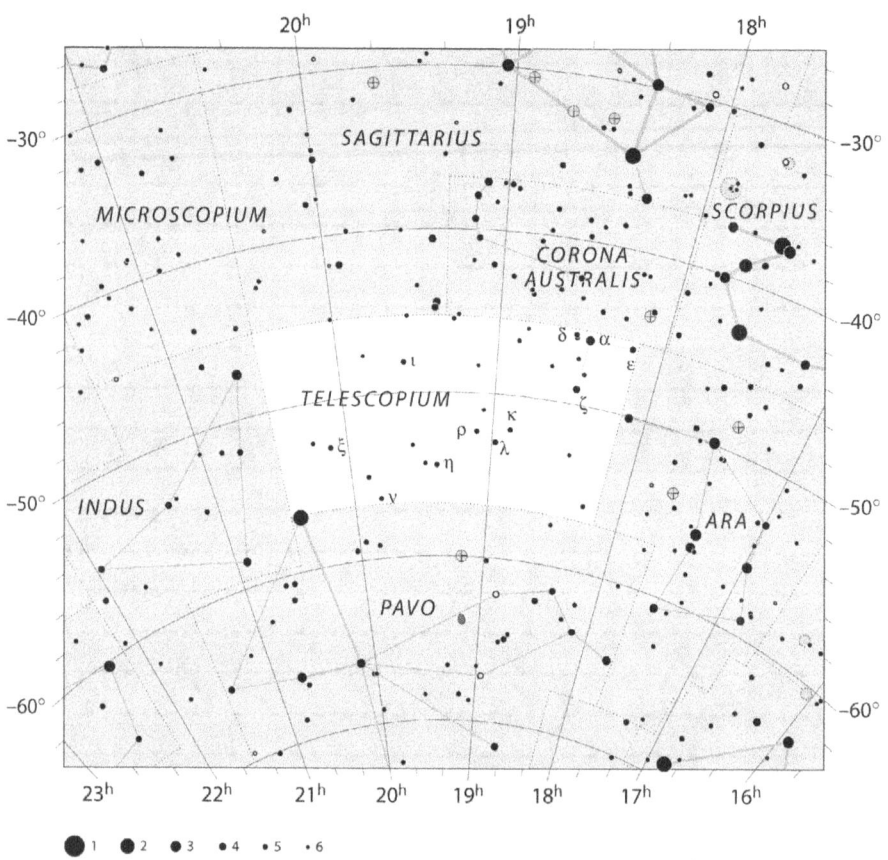

The Telescope

It is a shame that the most important instrument for astronomers should be associated with such a tiny portion of the sky, practically devoid of telescopic interest. Lacaille devised the constellation in the mid-eighteenth century.

Just east of the Scorpion's stinger lies the faint constellation of Corona Australis. To its south is found Telescopium. Its stars are mostly fourth and fifth magnitude.

Double stars:

This constellation has no outstanding binaries, and indeed all the visual binaries have rather faint primaries (sixth, seventh, or even eighth magnitude). For this reason none are listed.

Variable stars:

The brightest Mira-type variable is R Telescopii, which has a period of 462 days, and a range in magnitude of 7.8 to 14. The next maximum is expected in mid-March 1997; in 2000 the maximum should occur late in the year.

Deep Sky Objects:

The deep-sky objects in Telescopium -- mostly galaxies -- are all too faint for amateur study.

Triangulum

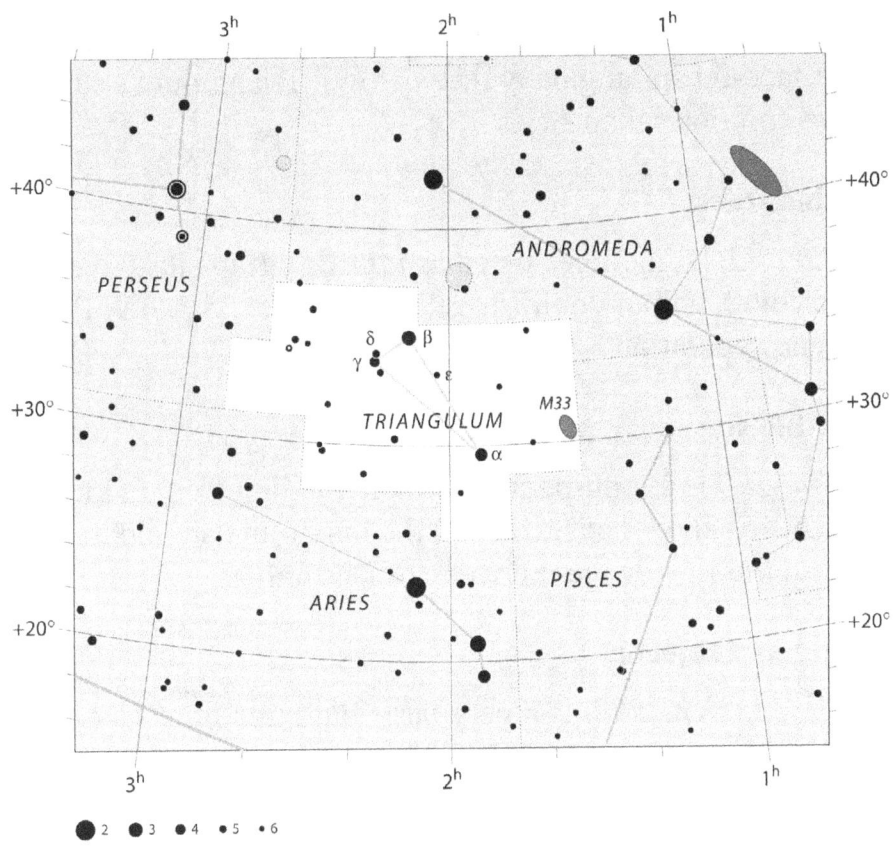

The Triangle

Triangulum lies just to the north of Aries. In antiquity its distinctive shape of three stars was called Deltoton.

Despite its small size it contains one of Messier's objects, a faint face-on spiral galaxy (see below). Triangulum's stars are even fainter than Aries'.

Double stars:

6 Trianguli [sometimes listed as *iota Tri*] is an attractive binary with colour contrast, yellow and blue: 5.3, 6.9; PA 71° and separation 3.9".

Variable stars:

R Trianguli is a long-period variable ranging from 5.4 to 12.6 every 266.9 days. In 2000 the maximum should occur around the first or second week of December.

Deep Sky Objects:

M33 (NGC 598) is a very large but quite faint face-on spiral galaxy sometimes known under the name "Pinwheel Galaxy" since it is said to be slowly rotating in a clockwise motion, making a complete turn probaby every 200 million years.

It's seventeen arc minutes west of alpha Trianguli and one degree north; or just about midpoint between alpha Arietis and beta Andromedae (slightly closer to the latter).

The galaxy is estimated to be from 2.5 to 3.5 million light years away. Low power scopes, or even binoculars, work best.

Triangulum Australe

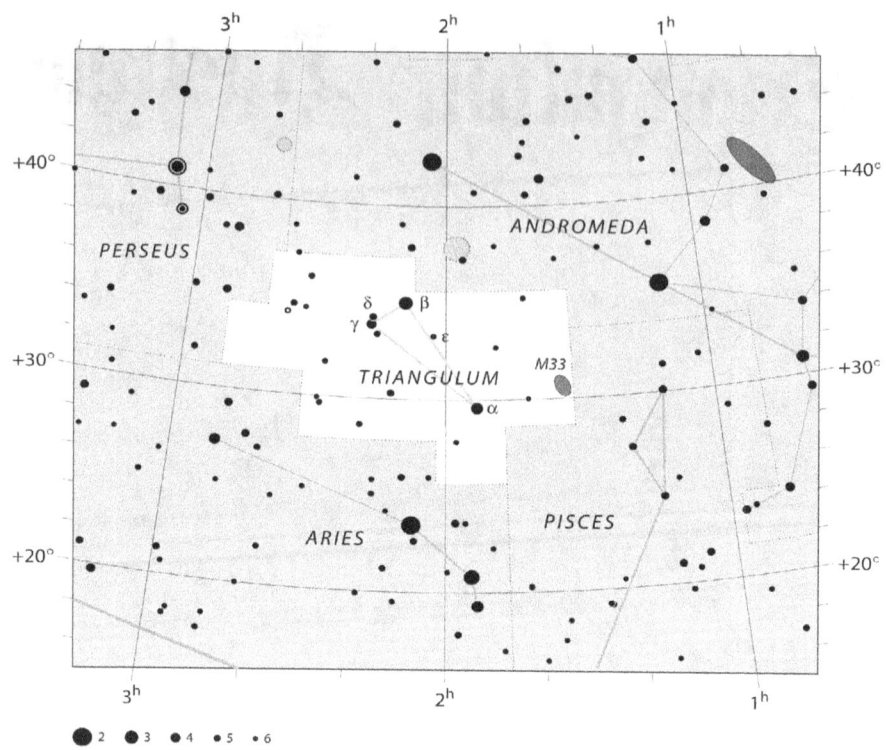

The Southern Triangle

Triangulum Australe, "The Southern Triangle", is one of the few constellations which has an obvious asterism. It was introduced by Johann Bayer in 1603.

The half-dozen Bayer stars range from 1.9 to 5.9 visual magnitude.

Double stars:

Triangulum Australe has none of any interest. *Iota TrA* is sometimes given as a binary (e.g. in Tirion's *Sky Atlas*) but other sources list this as optical only.

Variable stars:

R TrA is a cepheid varying from 6.0 to 6.8 every 3.4 days.

S TrA has a copper tint to it; this is a cepheid varying from 6.1 to 6.7 every 6.3 days. It's located less than one degree SE of beta TrA.

Deep Sky Objects:

NGC 6025 is a fairly bright open cluster of about thirty stars; it's found three degrees NNE of beta Trianguli Australis.

Tucana

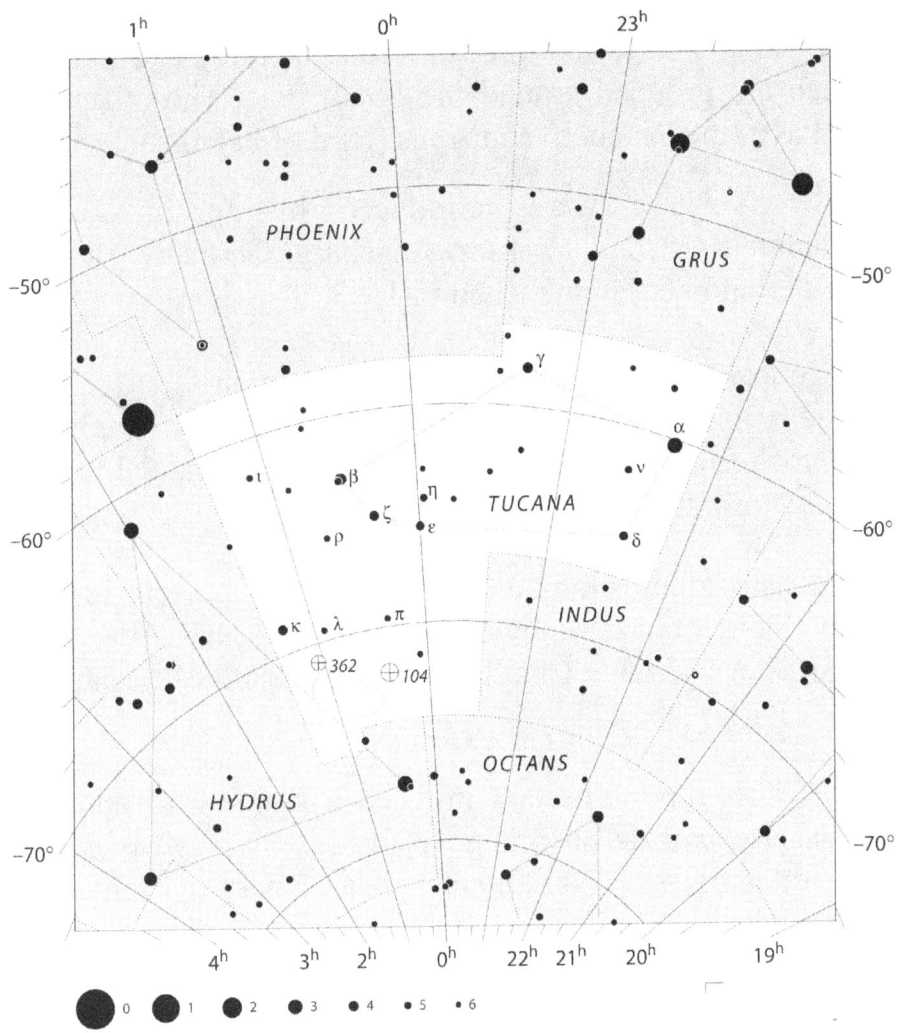

The Toucan

The toucan is a brightly colored bird of Central and South America, known for its over-large beak. Talkative and friendly, it makes a nice pet. Just feed it fruit and insects, along with the occasional lizard, and it will love you for life.

In the skies the Toucan is one of three exotic birds which are grouped around the South Pole. The other two are Pavo (the Peacock) and Apus (Bird of Paradise).

All three were introduced by Johann Bayer, an amateur astronomer from Augsburg, Germany. Bayer's book Uranometria, published in 1603.

The Toucan is one of the circumpolar southern constellations. If you live north of Mexico City or Bombay you won't find it. It has few Bayer stars, mostly at the four to five-magnitude range.

The constellation is famous for two deep sky objects: the large globular cluster known as 47 Tucanae (NGC 104), and the Small Magellanic Cloud, an unusual naked-eye galaxy (NGC 292).

The important stars in Tucana are few in number. *Alpha Tucanae* is an orange giant 130 light years away. Several other stars are of interest as binary systems (see below).

Double stars:

Beta^1 and *beta^2 Tucanae* form a splendid binary, part of a multiple system: 4.5, 4.8; PA 169°, 27".

Delta Tucanae has a faint companion: 4.8, 9.3; PA 282°, 6.9".

Kappa Tucanae: 5.1, 7.3; PA 336°, 5.4".

Lambda^1 Tucanae: 6.6, 8.0; PA 81°, 21".

Variable stars:

Tucana has a delta-Scuti type variable and a pulsating variable (Lb), neither one of which is of much interest to the amateur observer.

Theta Tucanae is a delta Scuti type variable with a range of 6.06 to 6.15, and period of one hour 11 minutes.

Nu Tucanae is a pulsating variable (Lb): 4.75-4.93 with uncertain period.

Deep Sky Objects:

Tucana is the home of the Small Magellanic Cloud and *47 Tucanae*, both of which are worthy of Messier.

47 Tucanae (NGC 104) is a splendid globular cluster, bright and large, a naked eye object which ranks alongside Omega Centauri. The cluster is about 20,000 light years away.

It is found 4° SW of lambda Tucanae.

The Small Magellanic Cloud (NGC 292) is an irregular galaxy, a cloud-like object perhaps 200,000 light years away.

The Small Magellanic Cloud is a companion to the Large Magellanic Cloud, in Dorado.

These Magellanic Clouds are actually neighbours of our own Milky Way Galaxy. They are so named because they were first noted by Ferdinand Magellan in 1519.

The term "irregular galaxy" refers to the fact that no apparent shape can be seen, and that a great amount of interstellar matter is visible. The cloud contains a large number of variable stars; well over a thousand have now been catalogued.

NGC 362 is another globular cluster, at the edge of the SMC but much closer (about 40,000 light years away). This cluster is nicely seen in binoculars.

NOTES

Ursa Major

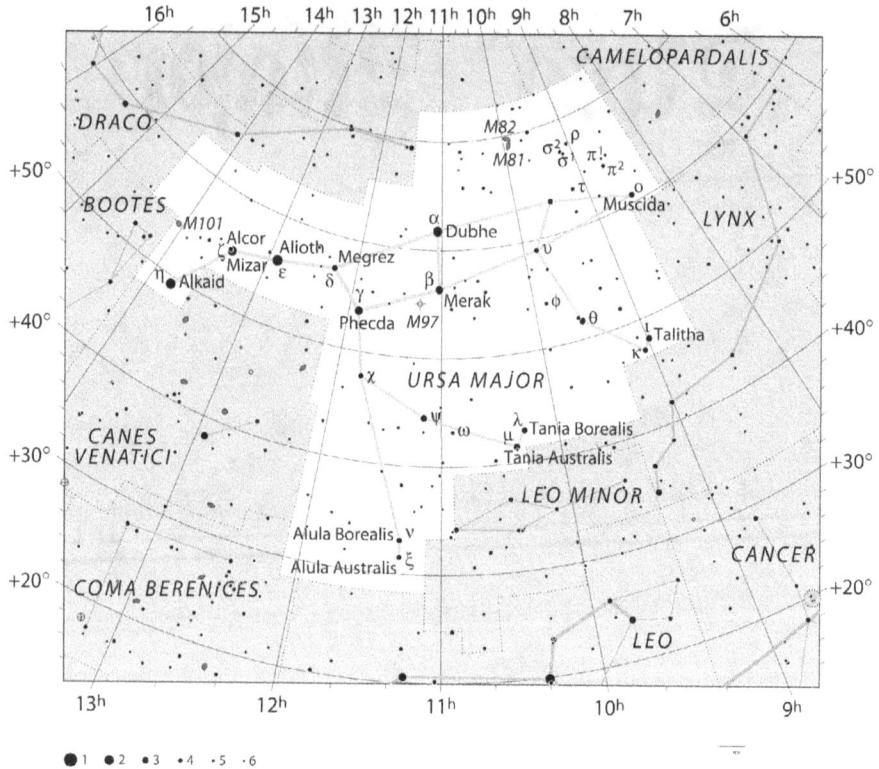

The Great Bear

The wood-nymph Callisto was a maiden in the wild region Arcadia. She was a huntress, "not one who spent her time in spinning soft fibers of wool, or in arranging her hair in different styles. She was one of Artemis' warriors, wearing her tunic pinned together with a brooch, her tresses carelessly caught back by a white ribbon, and carrying in her hand a light javelin or her bow" (Metamorphoses II 412-415).

Zeus caught sight of her and immediately desired her. He took on the shape of the goddess Artemis and spoke to Callisto, who was delighted to see the form of her mistress. She began to tell him of her hunting exploits, and he responded by raping her. "She resisted him as far as a woman could--had Hera seen her she would have been less cruel--but how could a girl overcome a man, and who could defeat Zeus? He had his way, and returned to the upper air" (Metamorphoses II 434-437).

The constellation itself doesn't have a particularly memorable shape to it; few of us bother to discover the form of a bear in the heavens. Rather it is the asterism known as the Big Dipper (or Plough if you live in the UK) that is most noticeable.

> The stars that make up the Big Dipper are seven in number, and follow the Greek alphabet, making them easy to remember. Thus, apart from *alpha Ursae Majoris*, the Dipper's bowl is made up of *beta*, *gamma*, and *delta*, then *epsilon*, *zeta*, and *eta* finish the asterism.
>
> This particular asterism has also a long history, seen in many cultures as a chariot or wagon. (Burnham, as one would expect, has a thorough discussion on this aspect of the constellation.)
>
> The seven are not moving in the same direction, and

over time the asterism will dissolve. In fact, it is only the last 50,000 years or so that a discernible "dipper" has formed. As the stars move their separate ways, the shape will more and more become plough-like, with the pointer star (*alpha Ursae Majoris*) moving in front of the rest, and somewhat south of its present position.

> This star, *alpha UMa* (Dubhe: the Bear), is a yellow giant, about 25 times the size of the Sun, and 86 light years away. It is a close visual binary, discussed below.
>
> *Beta Ursae Majoris* is named Merak, or "loin"; *gamma* is Phecda: thigh, and *delta* is called Megrez: root (or base) of the tail. These three are similar stars, all white (A-type) stars, and all within 100 light years distance.
>
> When we go out onto the tail, we first encounter *epsilon UMa*, an alpha-CV type variable (see below), and another A-type white star. Called Alioth (which no one has adequately translated) the star is one of the brightest in the constellation, although one of the more distant stars (if we go by its parallax). The star tables show a distance of only 64 light years. This distance is disputed and may be too small; the parallax indicates a distance of 360 light years.
>
> Then comes *zeta Ursae Majoris* (Mizar: girdle or apron). This star forms a fine binary (perhaps optical?) with *Alcor* (80 UMa) (see below). The star is 78 light years away according to its parallax.

Finally we have *eta Ursae Majoris*, called either Benetnasch or Alkaid, both of which mean "chief of the mourners". This is a blue-white star, a bit further than the rest at about 95 light years.

Other stars:

Omicron Ursae Majoris is called Muscida, meaning "muzzle", and indeed this star marks the bear's nose. Muscida has a very faint (15m) companion, a dwarf star at PA 191 degrees and separation of 7".

Nearby, marking the bear's ear, is a small group of stars made up of *sigma^1*, *sigma^2*, and *rho*. Our constellation map marks the spot simply with a lower-case sigma. The visual binary *sigma^2* is discussed below.

Xi Ursae Majoris is far to the south, marking one of the bear's feet. This star is not only an interesting binary, but also an historic one, as it was the first binary to have its orbit calculated (in 1828).

Double star:

Several have just been mentioned. But there are other binaries in Ursa Major worthy of investigating.

Dubhe (Alpha Ursae Majoris) is a well-known binary, with a close 4.8m companion which orbits every 44.66 years. In 2000 the values are: 1.9, 4.8; PA 214°, separation 0.6".

Phi UMa is even closer these days [PA 243°, separation 0.23"] but the distance is gradually widening. It has a period of 105.5 years. The two stars are similar in magnitude: 5.3, 5.4, resulting in a combined magnitude of 4.6.

Sigma² UMa is a much easier binary to resolve; presently the separation is 3.8" at PA 355°. The companion, a rather dim 8.2 visual magnitude, describes a leisurely 1067 year orbit. As with many slowly orbiting binaries, this one has had a variety of calculated periods, although Burnham's "best modern computation [of] 706 years" is now considered out of date.

Xi UMa is an attractive binary [4.3, 4.8] with a fast orbit. This star shouldn't cause too many problems to resolve; its closest point came in 1993 and it too is widening, presently sitting at PA 302° and separation 1.3". The star was designated an RS CVn type variable in 1993.

Zeta UMa, Mizar, is the best of the bunch and probably the easiest to find as well. A multiply system with Alcor, AB form a fixed binary at PA 152°, separation 14.4". Alcor (component C) is a distant 12 minutes east (709").

> Mizar was the first binary system to be discovered (in 1650), and is usually the first binary to be found and studied by amateur astronomers. No matter how long you study the stars, coming back to Mizar is always a treat.
>
> Both A and B are also spectroscopic binaries (that is, each one has a companion too faint for

observation but which shows up when studied spectroscopically). The presence of such a companion is deduced from changes in the doppler shift in the spectral lines of the primary.

Although at a great distance from Mizar (perhaps three light years away), *Alcor (80 UMa)* may be gravitationally bound to this star as it shares the same proper motion. However, most authorities believe the stars only form an optical binary.

This is a 3.99 visual magnitude star, 81 light years away. Alcor serves as a good jumping off point to study M101, a spectacular face-on spiral galaxy (see below).

Variable stars:

Ursa Major has no notable variables, but there are a number which might be of some interest.

Epsilon Ursae Majoris is an alpha-CVn type variable: 1.76-1.78 every five days and two hours.

> Alpha-Canum Venaticorum type stars are rotating variables which typically evince very little change in visual magnitude. These stars are generally A-type (that is, they have a spectrum range of B9-A5) but curiously enough they show an unusual abundance of a number of heavy metals and a corresponding lack in the more common elements.
>
> These stars are divided into three groups: those with predominantly silicon spectral lines, those with manganese, and those with chromium-strontium lines. Epsilon Ursae Majoris shows a

strong chromium line.

R Ursae Majoris is a Mira-type variable with period of 301.62 days, and a magnitude change from 6.5 to 13.7. Curiously, it is actually a brighter red when at its minimum; at maximum it loses much of its colour. The 2000 maximum was expected in the latter half of March.

Deep Sky Objects:

Ursa Major has five Messier objects: *M40, M81, M82, M97,* and *M101*.

M40 is the Messier object that really isn't one. In 1764 Messier went looking for an object that had been catalogued as a nebulosity in this area. What he found was two ninth-magnitude binary stars, very close together, which he assumed had been mistakenly catalogued as the nebulosity. However instead of leaving the matter there, he proceeded to catalogue the stars as his No. 40.

A hundred years later the stars were catalogued by Winnecke as binaries called "Winnecke 4"; they still go by this name. The binary (9.9, 9.3; PA 83 degrees, and separation 49") is found one and a half degrees north-east of *delta UMa*. The easiest way to find the binary is to draw a line from *delta* to 70 UMa, then another half a degree beyond this point.

M81 (NGC 3031) is a superb spiral galaxy, and with *M82* in the same field, half a degree to the north, forms a splendid pair.

The distance is approximately seven to nine million light years and, as Burnham reports, the galaxy is considered one of the most dense galaxies known, with a total mass of 250 billions suns. A large scope is needed to catch the fine detail in the spiral's arms.

M82 (NGC 3034) floats above M81 like an ethereal UFO; any minute you think it's going to zip away in the night sky.

The galaxy isn't, as one might think, a spiral on edge, but is usually described as spindle shaped. The galaxy is rather mysterious; it's thought that an explosion at its centre one and a half million years ago created the odd shape, which is still expanding at a rate of 950 km/second.

M97 (NGC 3587) often called the "Owl Nebula" for its two dark central areas (revealed only in the largest telescopes) resemble an owl's eyes. The nebula is formed by the still expanding shell of its central star, which is very small and compact, with a surface temperature as much as 85,000 kelvin.

M101 (NGC 5457) is a vast galaxy, one of the largest known, with open spirals. Although seen face on, it's fairly dim; it takes a large scope and an exceptionally good night to see this nebula at its best.

Located five and a half degrees east of *zeta UMa*, the usual method given to find M101 is to star hop. From *zeta UMa* (Mizar) proceed to Alcor, then over and slightly north to 81 UMa, and now down to the southeast to 83, then 84. Now locate 86 UMa, to the southeast. This star forms the bottom point of a wide-v shape with 84 and M101.

Some stars of note:

Groombridge 1830 and *Lalande 21185*, both of which require a finder's chart, and *47 Ursae Majoris*, which has recently been found to have a planet which could theoretically support water.

Groombridge 1830 is famous for having one of the largest proper motions, (7.050 arc seconds) third after Barnard's Star and Kapteyn's Star. In only 511 years it shifts its position by one degree. It's 28.8 light years away, and has a space velocity of 312 km/s. Its Epoch 2000 values are: right ascension 11h, 52m, 58.7s; declination +37 degrees, 43', 07".

Called a subdwarf, the star has about half the diameter of the Sun and only one-fifth of its luminosity; yet because it is so close, its visual magnitude is a fairly bright 6.45.

To locate Groombridge 1830 first draw a line between *phi Ursae Majoris* and *xi Ursae Majoris*. Move up this line to its midpoint then look to the east roughly at the same distance. You are now in the vicinity. There are several 6m stars in the region, but only one just north of a small but bright galaxy (NGC 3941). This is Groombridge 1830.

Lalande 21185 is a red dwarf of very small mass (about one third of the Sun's). Quite close at 8.2 light years, its proper motion is also very fast, at 4.777 arc seconds, and it has a space velocity of 187 km/s. Its Epoch 2000 values are: right ascension 11h, 03m, 20.2s; declination +35 degrees, 58', 13".

Lalande 21185 has a visual magnitude of only 7.49, and an absolute magnitude of 10.49. Finding the star could be an adventure; Burnham's finder's chart (on his page 1981) is useful.

First start from *xi UMa*. Star-hop to *nu UMa*, just to the north two degrees, then find the brightest star lying to the west (about six degrees). This is 46 Leonis Minoris; the fainter star just to the east is in Ursa Major, but is called 47 Leonis Minoris. Now move to the north one and three-quarter degrees. This is a rather bleak part of the sky. Two degrees to the east you're in the field; now you have to rely on the finder's chart.

NOTES

Ursa Minor

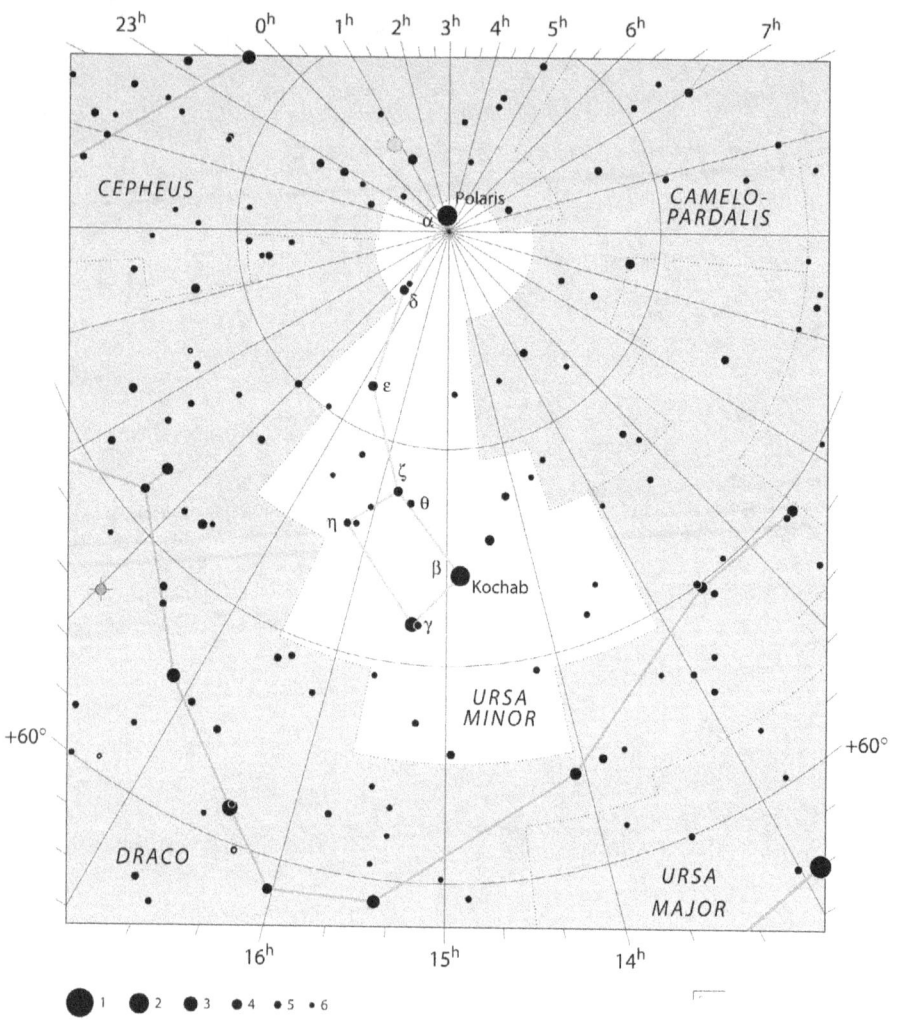

The Lesser Bear

Arcas was the son of Callisto, who was transformed by Hera into a bear. When Arcas was fifteen, he was out hunting in the forest when he came across a bear. The bear behaved quite strangely, looking him in the eyes. He of course could not recognize his mother in her strange shape, and was preparing to shoot her when Zeus prevented him. Arcas was transformed into a bear like his mother, and the two were taken up into the sky. Hera was annoyed that the pair should be given such honor, and took her revenge by convincing Poseidon to forbid them from bathing in the sea. It is for this reason that Ursa Major and Ursa Minor are both circumpolar constellations, never dipping beneath the horizon when viewed from northern latitudes.

Ursa Minor is better known as the Little Dipper.

Ursa Minor is a fainter version of the Big Dipper (or Plough, in the UK), and is home to the North Star.

The constellation dates from antiquity, and is said to have been introduced by the Greek philosopher Thales around 600 BC.

For as long as ships have sailed the seas Polaris has been an essential guiding star.

> However the Pole Star isn't, as one might think, constant, but rather it changes gradually through several thousands of years.
>
> The earth's axis moves very slowly, like a top, completing a circular path every 25,800 years. During this "precessional cycle" several stars take turns becoming the Pole Star.
>
> The present pole star, alpha Ursae Minoris, will be at the closest to the pole in 2102 AD, at which time it will

only be 27' 31" from the north pole. However beta UMi (Kochab), the brightest star in the constellation, is sometimes closer to the pole than is alpha, as it was 3000 years ago.

The brightest Pole Star is Vega (alpha Lyrae), which will resume this title in another 12,000 years. Another star which periodically becomes the Pole Star is Thuban (alpha Draconis), as it was some 4600 years ago, at the time of the pyramid building in Egypt.

Traditionally amateur astronomers have used the constellation as a rough guide on the clarity of the evening's sky. The stars range from second magnitude down to fifth (and even sixth); if these latter stars are clearly seen, it's a good night for viewing.

Ursa Minor has one notable binary and a few variables.

Double stars:

Alpha UMi is a well-known double star with a wide ninth magnitude companion: 2.1, 9.1; PA 218°, separation 18.4".

Variable stars:

Alpha UMi (Pole Star) is a cepheid varying from 1.92 to 2.07 every 3d 23h 16m 28.8s.

Gamma UMi is a delta Scuti type variable with extremely small range (3.04 - 3.09) every 3h 26m.

Epsilon UMi is an EA type variable: 4.19 - 4.23, period 39.48d

NOTES

Vela

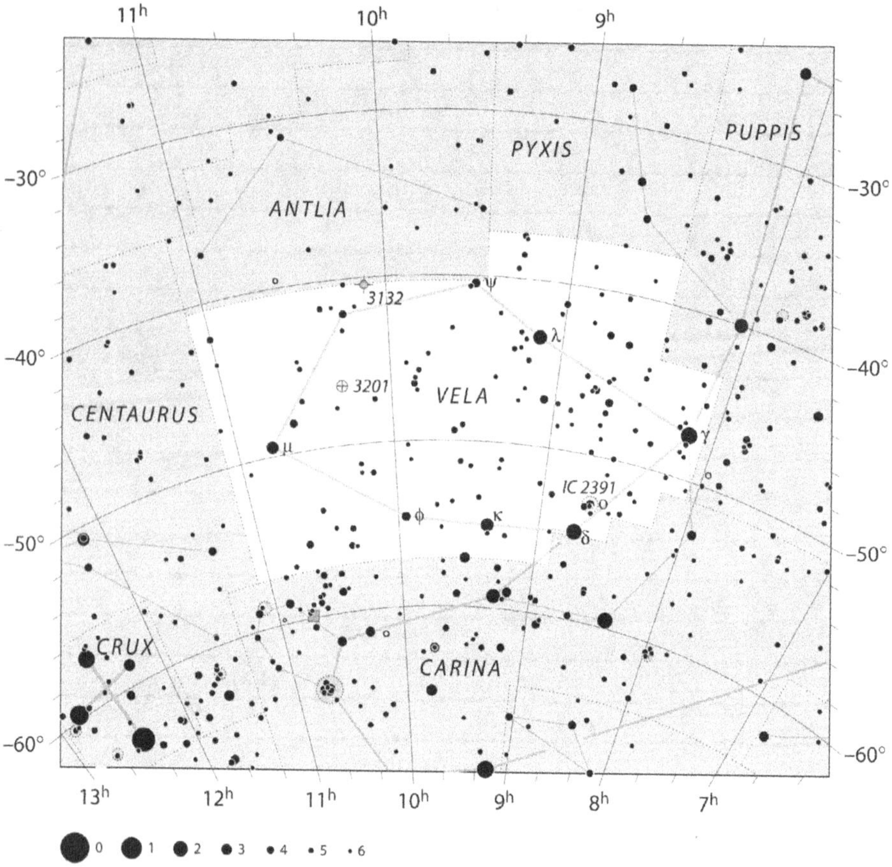

The Sail

As mentioned in regard to "Carina", Jason and his Argonauts sailed off in the Argo Navis to capture the Golden Fleece. The constellation that commemorated that adventure is now broken up into three smaller constellations: Carina (the Keel), Puppis (the Stern) and Vela (the Sail).

The stars that make up the sail are widely dispersed. Some cartographers of the night sky add a few more stars to make the sail more billowy. Since Vela is part of the original Argo Navis, it has only a few of the Bayer stars of that larger constellation.

Sometimes observers associate the lower two stars, kappa Velorum and delta Velorum, with iota Carinae and epsilon Carinae, and believe they are looking at the Southern Cross. The real Southern Cross is in the nearby constellation of Crux; this cross shared by Vela and Carina goes by the name of the False Cross.

The stars that make up the sail are widely dispersed. Some cartographers of the night sky add a few more stars to make the sail more billowy. Since Vela is part of the original Argo Navis, it has only a few of the Bayer stars of that larger constellation.

Sometimes observers associate the lower two stars, kappa Velorum and delta Velorum, with iota Carinae and epsilon Carinae, and believe they are looking at the Southern Cross. The real Southern Cross is in the nearby constellation of Crux; this cross shared by Vela and Carina goes by the name of the False Cross.

Although Vela does not make much of a sight in the southern skies, it does have a number of notable objects, including the brightest Wolf-Rayet star, an optical pulsar, and a pulsating variable which is the prototype of an entire class of cepheids.

Double stars:

Gamma Velorum is not only a fixed double (AB), but there are two other wide components. This is also a notable *Wolf-Rayet* star (see below).

> AB: 2.2, 4.4; PA 220°, 41.2";
> AC: 8.5, 151°, 62.3";
> D: 9.4, 141°, 94".

Delta Velorum is a multiple system as well, with component C having its own companion.

> AB: 2.1, 5.1; PA 153°, 2.6";
> AC: 10.5, PA 61°, 69.5";
> Cc: 10.5, 13; PA 102°, 6.2".

Psi Velorum is the most interesting binary in Vela. It is a close visual binary with a very rapid orbit. Presently the values are: PA 214°, separation 0.48".

Variable stars:

Lambda Velorum is a supergiant irregular variable which changes slightly in magnitude from 2.14 to 2.3.

Y Velorum is a Mira-type variable, from 8.0 to 14.2, every 444.61 days. In the year 2000 the maximum should occur around the second of January.

AI Velorum is a notable pulsating variable now grouped into a select number called *dwarf cepheids*. This star is the brightest of the 70 or so known dwarf cepheids, varying from 6.4 to 7.1, bright enough to be classified by the Webb Society as a "binocular variable".

Dwarf cepheids typically have a period of from

2.4 hours to 4.8 hours (i.e. 0.1-0.2 days), although as a group they range from as short as 1h20m to as long as 6h. *AI Velorum*'s period is 2h40m.

Pulsating variables change in visual magnitude due to sporadic movement in their outer layers. The pulsation occurs when there is an imbalance between gravitational pull (inward) and gaseous pressure (outward), causing a continuous cycle of expansion-contraction. [See Valerie Illingworth, *Facts on File Dictionary of Astronomy* for a detailed explanation.]

While the brightest of the known dwarf cepheids, AI is not that easy to find. First locate gamma Velorum, then move up three degrees where you find a fairly bright but unnamed star (HD 68217). In the same view you should see AI to the southeast one degree. If this method proves difficult, you might try the finder's chart in Burnham, p. 2039.

Gamma² Velorum (and Wolf-Rayet stars in general)

Wolf-Rayets form a very rare type of star having extremely hot surfaces (perhaps up to 90,000 kelvin) and ejecting gas: WC eject predominately helium and carbon, WN nitrogen and helium, and the even more rare WO stars eject oxygen.

Named after C.J.E. Wolf and G. Rayet, who discovered the existence of this type of star in 1867 (at the Paris Observatory), these stars are essentially left-over centres of giant O stars, which have ejected their helium and nitrogen atmospheres. These stars are typically around ten solar masses, and many of them are binaries (such as gamma Velorum).

Most Wolf-Rayets are quite distant from us; gamma Velorum is the closest known W-R star at an estimated 550-800 light years, although some catalogues list a greater distance.

Deep Sky Objects:

Vela has quite a few deep sky objects, but none could be described as outstanding.

IC 2391 is an open cluster of ten or so stars including omicron Velorum. This is a rather bright and large scattered cluster considered to be about 500 light years away.

IC 2395 is another open cluster, much smaller and considerably fainter (it's about 3000 light years distant) but it contains about twice as many stars as IC 2391.

The cluster is nearly six degrees east of gamma Velorum, in a particularly dense part of the Milky Way. It might be easier to find the most prominent star in the area first.

This star (*b Velorum*), 3.9m, is a bright binary (companion is a 10m star, very wide separation: 37.5"). It's five degrees east of gamma Velorum, and about one degree north, in the middle of a particularly rich area of the sky, but clearly the brightest star in the area.

Now star-hop south. First, at about one degree south is

n Velorum (4.77m). Then at two degrees south of *b Velorum* you'll find a rather dimmer star (HX Vel, 5.5m, a variable and binary). This star is probably not a part of the IC 2395 cluster as its distance is considerably closer than the cluster's).

In the same viewing area as HX Velorum, about half a degree east, is IC 2395. You'll see another cluster in the same field, to the southeast: NGC 2670, which is similar in size and brightness.

NGC 3132 is a planetary nebula, and the best deep sky object in Vela.

> This planetary nebula is found in a rather isolated part of the sky, about six degrees east of psi Velorum, just inside the border with Antlia. The closest bright star is *q Velorum* (3.85m), two degrees to the southeast.
>
> NGC 3132 has a fairly bright 10m central star which (as Burnham points out) isn't really the star which is furnishing the nebula's illumination. This light comes from a 16m dwarf companion (separation 1.65") which has a very hot surface, about 100,000 kelvin.
>
> Unlike most planetary nebulae, this one isn't round, but is more oblong, which gives it the name it sometimes goes by: the Eight-Burst Nebula. It is estimated at from 2000-3000 light years away. It's about the same size as the Ring Nebula in Lyra, and almost as bright.

Virgo

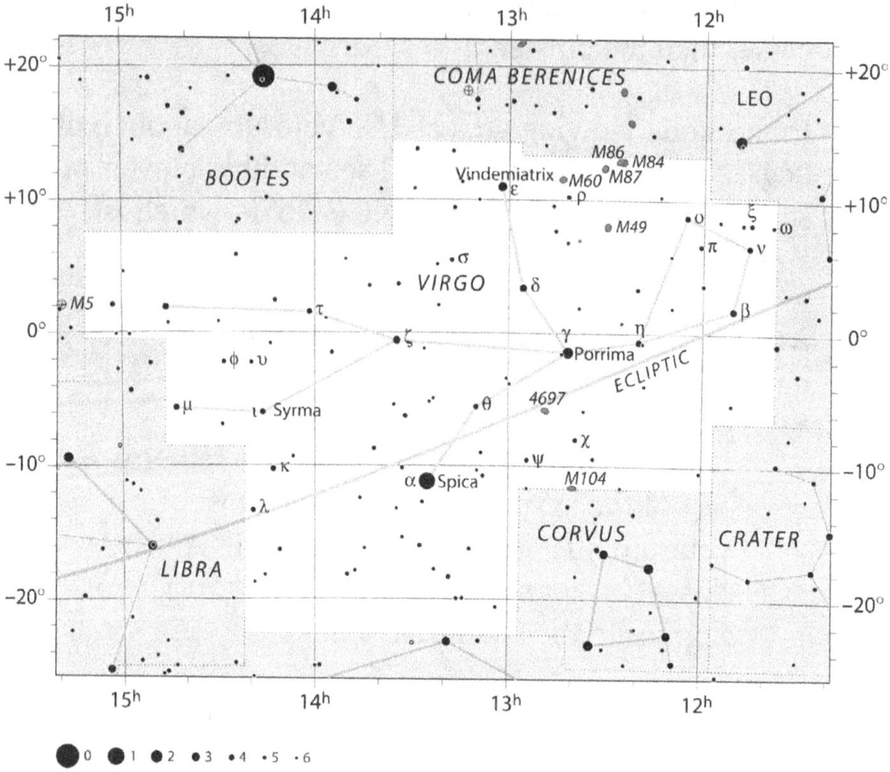

The Virgin

Virgo is a zodiacal constellation. According to the ancient poets, the virgin is also sometimes known as Astraea. She lived on the earth during the Golden Age of man, which is described by Hesiod:

> *First a golden race of mortal men were*
> *made by the immortals who have Olympian homes.*
> *They lived in Kronos' [Saturn's] time, when he ruled the sky,*
> *they lived like gods, with carefree heart,*
> *free and apart from trouble and pain; grim old age*
> *did not afflict them, but with arms and legs always*
> *strong they played in delight, apart from all evils;*
> *They died as if subdued by sleep; and all good*
> *things were theirs; the fertile earth produced fruit*
> *by itself, abundantly and unforced; willingly and*
> *effortlessly they ruled their lands with many goods. But*
> *since the earth hid this race below,*
> *they are daimones by the plans of great Zeus*
> *[Zeus], benevolent earthly guardians of mortal men,*
> *who watch over judgments and cruel*
> *deeds, clothed in air and roaming over all*
> *the earth (Works and Days 109-125).*

The "daimones" of which Hesiod speaks are invisible spirits which watch over men. Presumably, although it is unclear, Astraea is the daimone whose province is justice. The emblem of her office was therefore the scales (Libra), which are seen next to Virgo in the sky.

Virgo is the second largest constellation and is

highest in the northern hemisphere during May and June. The brightest star in Virgo is Spica.

Mythology, of course, influenced the naming of many objects in the night sky, not just the constellations. The planets all bear names from Roman mythology which reflect their characteristics: Mercury, named for the speedy messenger god, revolves fastest around the sun; Venus, named for the goddess of love and beauty, shines most brightly; Mars, named for the god of war, appears blood-red; Zeus, named for the single most important god, is the largest planet in our solar system. Even the names of the Galilean moons of Zeus (the four largest, which may be seen with even a small telescope) are drawn from mythology. Io, Europa, Ganymede, and Callisto were all desired--and taken by force--by Zeus. It is ironic that the mythological characters mythological women the king of the gods so ardently pursued now revolve around him.

Virgo is unique in that it is the only constellation containing all the Bayer stars with no additional superscript letters or numbers: just the Greek alphabet from alpha to omega.

Alpha Virginis is known as *Spica*: the "ear of wheat" that the goddess is carrying.

Spica is a blue-white eclipsing binary with a period of just over four days. The star is about twice the size of the Sun, but with a luminosity of about 2000 times the Sun.

Gamma Virginis carries the name of the Roman goddess of prophecy: *Porrima*.

Porrima is a notable binary of twin stars (see below). It's 32.9 light years distant and has the diameter of 1.5 Suns.

Double stars:

Gamma Virginis is a splendid binary of similar 3.5 magnitude stars, with a recently revised orbit of 168.8 years. The 2000.0 values are PA 260° and separation 1.5".

Theta Virginis is a white star with two companions, both rather faint: AB: 4.4, 9.4; PA 343, separation 7.1"; AC: 4.4, 10.4; PA 298°, separation 70".

Phi Virginis is a fixed binary: 4.8, 9.3; PA 110°, separation 4.8". The primary is a delicate yellow.

Struve 1719 is a striking binary of nearly equal stars: 7.3, 7.8; PA 1°, separation 7.5".

The star is located exactly midway between *zeta* and *gamma Virginis*, north about two degrees from a line joining these two stars. Another way to find it would be to form a triangle with *zeta, gamma,* and *delta Virginis*. The star is at the centre of this triangle.

Struve 1833 is even more attractive: 7.0, 7.0; PA 172°, separation 5.7".

This system is located 2.5° SE of *iota Virginis*. If using Tirion's *SkyAtlas*, you'll find two binaries in this region. Struve 1833 is the northern one. (The other is a triple system called β939. See Burnham for its details.)

Struve 1869 is the third of our trio of Struve binaries. Another lovely sight, but a bit of a challenge: 8.0, 9.0; PA 133°, separation 26".

To find this one, move southeast of mu Virginis two degrees.

Variable stars:

A number of stars show very little variablility, such as *alpha Virginis*, an "Ell." type variable: 0.95 to 1.05 ever four days and *rho Virginis* is a delta Scuti variable: 4.86-4.88.

R Virginis is a long-period variable with a range from 6.2 to 12.1 every 145.63 days, exceptionally short for a Mira type variable. In 2000 the maximum should occur in the first week of June.

Deep Sky Objects:

Virgo has some exceptional deep sky objects: the Virgo Galaxy Cluster, which contains eleven Messier Objects, more than any other constellation except Sagittarius (which has 15). There are also many fine NGC objects in the same vicinity, some just as splendid as the Messiers (such as NGC 5364 and the Siamese Twins: NGC 4567 and 4568).

Then there is the quasar *3 C 273*, thought to be from two to three billion light years away.

The region from Coma Berenices down through Virgo is renowned for its galaxies: the Virgo Galaxy Cluster, considered to be about 42 million light years distant. In the midst of dozens of bright galaxies are eleven chosen by Messier for his catalogue.

It is impossible to is give specific directions to locate each Messier object in an extremely rich field. Burnham (p. 2075) gives a useful grid to assist in their location, and recommends at least a six inch telescope. You may find that an even larger scope is necessary to get the most out of this region.

M49: a bright elliptical found between two six magnitude stars.

M58: bright compact barred spiral, but it takes a good night and at least a medium sized telescope to see the central bar.

> The Siamese Twins (NGC 4567 and NGC 4568) are 0.5 degree southwest: two faint galaxies seemingly joined in the middle.

Also in the same vicinity are *M59* and *M60*: two small but bright ellipticals.

M61: armed spiral seen face-on, very bright. This is one of the largest galaxies associated with the Virgo Cluster, and may have a mass of fifty billion Suns. Three supernovae have occurred in M61, the last in 1964.

M84, M86, and *M87*: three more ellipticals, in a very rich region. M87 is the centre of the Virgo Cluster, and is one of the most luminous galaxies known.

M89: small elliptical, resembling M87 but fainter.

M90: nice spiral in same region as M89.

M104: The Sombrero Galaxy. Truly magnificent, this galaxy is isolated from the rest (although apparently is still a member of the Virgo Cluster).

> *Seen edge-on, the huge luminous nucleus is surrounded by a dark dust lane, which should be visible even in smaller telescopes (depending on the quality of the night sky).*

The quasar *3 C 273* has a variable magnitude, roughly from 12 to 13. Its exact 2000 epoch location is: right ascension 12h29.1m, declination +2 degrees, 3.2'; or about 3.5 degrees northeast of *eta Virginis*.

The name comes from "quasi-stellar object". A single quasar can emit more energy than a hundred galaxies, emitted (in the most part) in the form of infared radiation.

It was this object in Virgo, *3 C 273*, that was first identified as a non-stellar object, by Maarten Schmidt, from the analysis of its redshift.

Quasars are perhaps the most luminous object known; some (including 3 C 273) are known to have absolute magnitudes as great as -27.

> *Burnham (p. 2101) gives an identification chart, as well as a detailed discussion on the phenomenon of quasars.*

NOTES

Volans

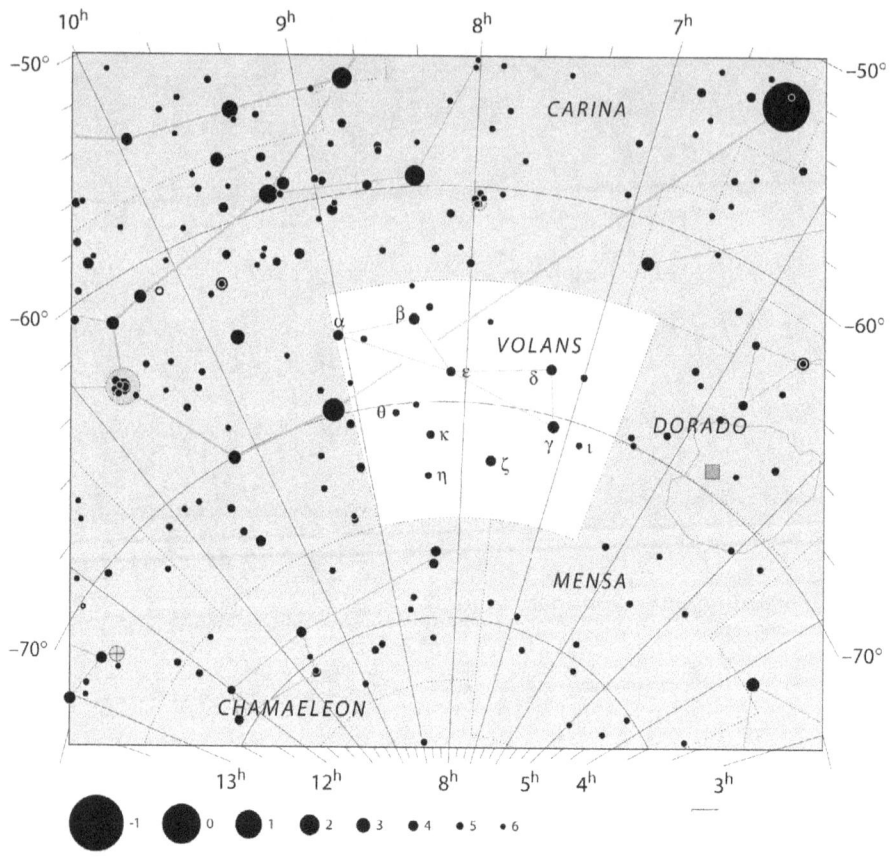

The Flying Fish

Volans is one of those constellations introduced by Johann Bayer in his 1603 star atlas. He called it Piscis

Volans; only the adjective has survived. The asterism shows a sideways view of the "flying fish" (sort of).

This is one of the few constellations in which alpha is not the brightest star; *beta Volans* is slightly brighter, and three other stars are brighter than alpha Vol: Bayer stars.

Volans has few attractions for the amateur observer: two fine binaries and a faint galaxy.

Double stars:

Gamma^2 and *gamma^1 Volantis* form a fine binary, a deep yellow primary and whitish component: 3.9, 5.8; PA 300°, separation 13.6". Notice that gamma^2 is the primary.

Epsilon Volantis has a faint component: 4.5, 8.1; PA 24°, separation 6.1".

Variable stars:

Volans has no variables suitable for amateur observation.

Deep Sky Objects:

There is one galaxy here which may be of some interest, a fairly faint but fine example of a barred spiral galaxy.

NGC 2442 is a faint barred spiral galaxy found midway between gamma and epsilon Volantis.

Vulpecula

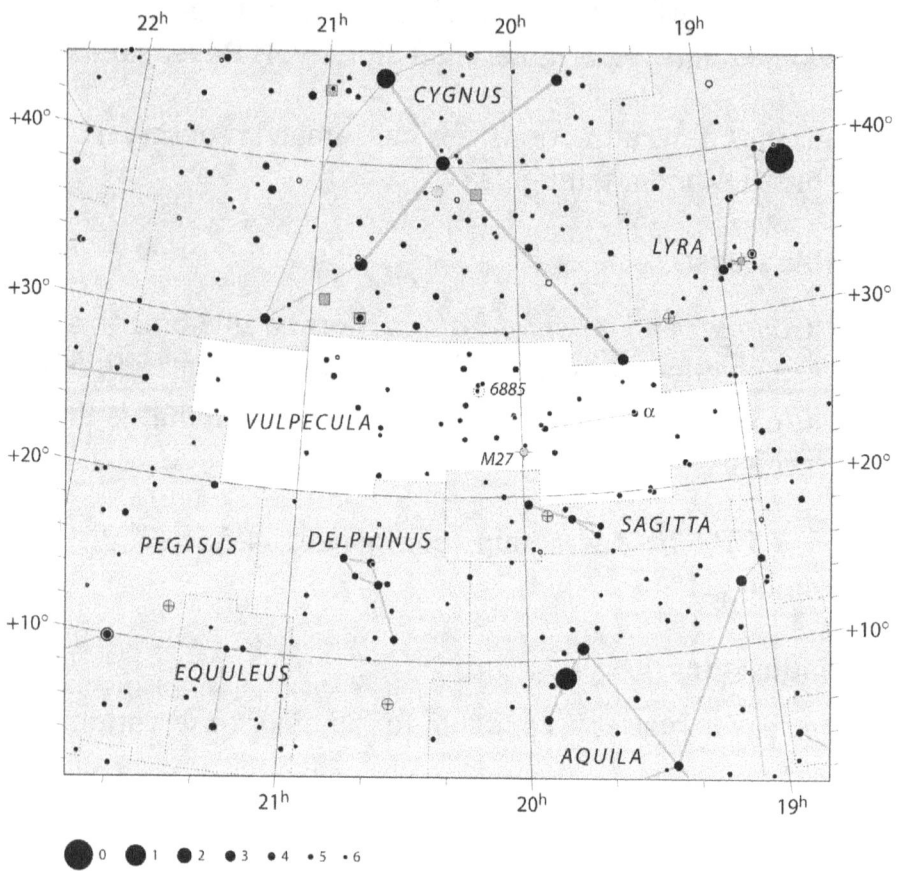

The Fox

Vulpecula, The Fox, was originally called Vulpecula cum Anser, The Fox and Goose.

Vulpecula has several objects of interest: a fine binary, a couple of variables, and even a Messier object.

Double stars:

Alpha Vulpeculae may be only optical (observers differ on this point). AB: 4.6, 6.0; PA 28°, separation 13.7".
16 Vulpeculae is a close binary with nearly equal components: 5.9, 6.3; PA 115°, separation 0.8".

Struve 2525 is a fine binary with orbit of 990 years; the 2000 values are: 8.5, 8.7; PA 291°, 2.1". The binary lies between beta Cygni and 3 Vulpeculae.

Variable stars:

R Vulpeculae is a Mira type variable with range of 7.4 to 13.7 every 137 days.

Deep Sky Objects:

M27 (NGC 6853), "The Dumbbell Nebula" is a noted planetary nebula, large, bright, and oddly shaped (thus its name). It glows with a faintly green colour.
    The nebula is found midway between 12 Vul and 17 Vul and about half a degree to the south. (14 Vul is in the same field, just to the NNW).

NGC 6940 is an open cluster of about a hundred stars, found just midway between 23 Vul and 32 Vul, and a half degree to the north.

Glossary

Apparent magnitude: the apparent brightness of a celestial object as observed from Earth

Asterism: a prominent pattern or group of stars, typically having a popular name but smaller than a constellation

Bayer star: a stellar designation in which a specific star is identified by a Greek letter, followed by the genitive form of its parent constellation's Latin name. The original list of Bayer designations contained 1,564 stars.

Binary star: a star system consisting of two stars orbiting around their common center of mass.

Double star: a pair of stars that appear close to each other in the sky as seen from Earth when viewed through an optical telescope.

Magnitude: a measure of brightness and brightness differences.

Messier objects: a set of astronomical objects first listed by French astronomer Charles Messier in 1771

Orbit: the curved path of a celestial object or spacecraft around a star, planet, or moon, esp. a periodic elliptical revolution.

Proper Motion: a stars angular change in position over time as seen from the center of mass of the Solar System. It is measured in seconds of arc per year, arcsec/yr, where 3600 arcseconds equal one degree.

Bibliography

Allen, Richard H. (1899). *Star Names: Their Lore and Meaning*. G. E. Stechert. OCLC 30773662.

Bakich, Michael E. (1995). *The Cambridge Guide to the Constellations*. Cambridge University Press. ISBN 978-0-521-44921-2.

Davis, George A., Jr. (1944). "The Pronunciations, Derivations, and Meanings of a Selected List of Star Names". *Popular Science* **52**: 8. Bibcode:1944PA.....52....8D.

French, Sue (January 2006). "Winter wonders: star-studded January skies offer deep-sky treats for every size telescope". *Sky and Telescope* (Academic OneFile) **111** (1): 83. (subscription required)

Higgins, David (November 2002). "Exploring the depths of Andromeda". *Astronomy* (Academic OneFile): 88.

Koch, Andreas; Grebel, Eva K. (March 2006). "The Anisotropic Distribution of M31 Satellite Galaxies: A Polar Great Plane of Early-type Companions". *Astronomical Journal* **131** (3): 1405–1415. arXiv:astro-ph/0509258. Bibcode:2005astro.ph..9258K. doi:10.1086/499534.

Hoskin, Michael; Dewhirst, David (1999). *The Cambridge Concise History of Astronomy*. Cambridge University Press. ISBN 978-0-521-57291-0.

Jenniskens, Peter (2006). *Meteor Showers and Their Parent Comets*. Cambridge University Press. ISBN 978-0-521-85349-1.

Makemson, Maud Worcester (1941). *The Morning Star Rises: an account of Polynesian astronomy*. Yale University Press.

Moore, Patrick; Tirion, Wil (1997). *Cambridge Guide to Stars and Planets* (2nd ed.). Cambridge University Press. ISBN 978-0-521-58582-8.

Moore, Patrick (2000). *The Data Book of Astronomy*. Institute of Physics Publishing. ISBN 978-0-7503-0620-1.

Olcott, William Tyler (2004). *Star Lore: Myths, Legends, and Facts*. Courier Dover Publications. ISBN 978-0-486-43581-7.

Pasachoff, Jay M. (2000). *A Field Guide to the Stars and Planets* (4th ed.). Houghton Mifflin. ISBN 978-0-395-93431-9.

Rao, Joe (October 2011). "Skylog". *Natural History* (Academic OneFile) **119** (9): 42. (subscription required)

Ridpath, Ian; Tirion, Wil (2009). *The Monthly Sky Guide* (8th ed.). Cambridge University Press. ISBN 978-0-521-13369-2.

Ridpath, Ian (2001). *Stars and Planets Guide*. Princeton University Press. ISBN 978-0-691-08913-3.

Rogers, John H. (1998). "Origins of the Ancient Constellations: II. The Mediterranean Traditions". *Journal of the British Astronomical Association* **108** (2): 79–89. Bibcode:1998JBAA..108...79R.

Russell, Henry Norris (October 1922). "The new international symbols for the constellations". *Popular Astronomy* **30**: 469. Bibcode:1922PA.....30..469R.

Sherrod, P. Clay; Koed, Thomas L. (2003). *A Complete Manual of Amateur Astronomy: Tools and Techniques for Astronomical Observations*. Dover Publications. ISBN 978-0-486-42820-8.

Staal, Julius D.W. (1988). *The New Patterns in the Sky: Myths and Legends of the Stars* (2nd ed.). The McDonald and Woodward Publishing Company. ISBN 978-0-939923-04-5.

Thompson, Robert Bruce; Thompson, Barbara Fritchman (2007). *Illustrated Guide to Astronomical Wonders*. O'Reilly Media. ISBN 978-0-596-52685-6.

Wagman, Morton (2003). *Lost Stars*. McDonald and Woodward Publishing. ISBN 978-0-939923-78-6.

Wilkins, Jamie; Dunn, Robert (2006). *300 Astronomical Objects: A Visual Reference to the Universe* (1st ed.). Firefly Books. ISBN 978-1-55407-173.

Charles Messier's Catalog of Nebulae and Star Clusters". SEDS. 15 June 2007. Retrieved 2010-05-08.

Coder, Errol Jud (2013). *The Constellations: Myths of the Stars*. Dreamscape Publishing. ISBN 978-1483999609 .

Other books by
Errol Jud Coder

The Constellations: Myths of the Stars
2nd Edition, 2013

www.ingramcontent.com/pod-product-compliance
Lightning Source LLC
Chambersburg PA
CBHW071353170526
45165CB00001B/28